Introduction

In this book I want to debunk the most expensive hoax ever perpetrated on mankind that has so far cost £billions and is going to cost £trillions if it continues along its present path. This is money that could have been better spent on alleviating poverty, researching treatments for cancers and other diseases and a multitude of other worthwhile causes. Instead it has lined the pockets of the liars and fraudsters who seek to perpetuate the scam and gain academic credibility on a subject to which they are not entitled together with political power, to which they are even less entitled.

I have written this book with the intention of exposing the lies, fraud, corruption, political motivation and greed that have been at the heart of this scam. I am talking of course about 'Global Warming', which morphed into 'Anthropogenic Global Warming' (AGW) with its final mutation being 'Climate Change', following an unexplained pause in warming of over 18 years. Any term where the syntax has to be changed twice, to accentuate its impact, cannot possibly be credible, especially when the second change was made, because there was no warming, global or otherwise.

My intention is to present easy to understand facts even for those readers who have no scientific knowledge or inclination. This book is not scientific it is a treatise (a systematic exposition or argument in writing including a methodical discussion of the facts and principles involved and conclusions reached). Thanks to the Merriam Webster Dictionary for this definition. My aim is to dispel a myth whose purpose is much more sinister than that of saving the planet. I hope to do this by informing and entertaining my readers at the same time. I would like to thank you for buying, borrowing, or begging a copy of this book, but more importantly for reading it and lobbying politicians to act on its content.

One thing I need to state at the very beginning is that all quotations, references to published papers, excerpts from books etc are all paraphrased. I have had to re-write the original book due to the publishers checking it for plagiarism and refusing to publish it in its previous form. I accept and defend the rights of those whose words are effectively stolen to provide material for a book, magazine or any other written or spoken work. Theft of words to make a profit is no different to theft of goods or services. However, when the 'plagiarised' work has quotation marks, the attributes of its source and a reason for its inclusion either to support or oppose a premise, then it prevents a reasoned argument and debate from both sides.

I have no problem at all with anyone copying and pasting any part of this book into a publication, whether in agreement or disagreement with my words, provided that the source is also credited. In one of the appendices I wanted to copy United Nations Agenda 2030, which I presumed would be in the public domain and probably is, but the plagiarism publishing rules preclude its incorporation into this book. Since this is an important part of the book I intend to amend the e-book by adding a hyperlink to my responses to some of its content in the near future, but I need to sort out any copyright issues first. When dealing with a large bureaucratic organisation this will not be an easy task and I anticipate it taking a long time. The subject of perceived plagiarism and the consequent impossibility of having a reasoned debate could be the subject of another book.

Chapter 1: A Brief Synopsis.

When the Earth first came into being, the atmosphere consisted of 20% carbon dioxide this compares to 0.04% currently, an increase from 0.032% in 1950 and 0.028% in the 18th Century. The Sun was 30% less luminous when Earth was first formed than it is now, but according to 'climatologists' the 200+ year minute increase in atmospheric carbon dioxide is causing a massive change in Earth's climate. I do not believe this for one second and I hope those who read this book realise that the whole concept of 'human induced climate change' is wrong and wholly politically contrived.

It is 10 years since the then disastrous Brown government passed the UK Climate Change Act with its legally binding commitment to ensure that CO2 emissions in 2050 are at least 80 per cent lower than in 1990. This has caused energy prices to skyrocket, together with more spent on 'Smart Meters' allowing energy companies to increase the unit cost of energy on an hourly basis to reduce energy consumption. Of course those who voted for this Act of Parliament thought that people would regard power cuts and high energy prices as necessary to save the planet. This is how the whole 'Globull' (as in bullsh*t) Warming farce started; the first three words of the next paragraph are the key as will be revealed later.

The United Nations Framework Convention on Climate Change (UNFCCC) is an international environmental treaty adopted on the 9th of May 1992. It was formalised at the Earth Summit in Rio de Janeiro from the 3rd to the 14th of June 1992. After it was thought enough countries had ratified it, it then entered into force on the 21st of March 1994. The UNFCCC objective is to 'stabilise greenhouse gas concentrations in the atmosphere at a level that would prevent dangerous anthropogenic interference with the climate system'. The framework sets non- binding limits on carbon dioxide emissions for individual countries but does not contain any enforcement mechanisms. Instead, the framework outlines how specific international protocols or agreements (treaties) may be negotiated to specify further action towards this objective.

Initially, an Intergovernmental Negotiating Committee (INC) produced the text of this Framework Convention during its meeting in New York from 30th of April to 9th of May 1992. The UNFCCC was adopted on 9th of May 1992, and opened for signature on 4th of June 1992. The UNFCCC has 197 participating countries as of December 2015. The convention enjoys broad legitimacy, largely due to its nearly universal membership (their words, not mine). The problem is the science is politically inspired, is basically flawed and if implemented in full, it will mean that mankind will be floored too!

This crap was written and believed despite well established knowledge that atmospheric concentrations of carbon dioxide in the past were up to 500 times higher than they are now when life on Earth did not just exist, it thrived . Carbon dioxide is plant food, plants are at the base of the food chain, more carbon dioxide means more plant growth, which means more people and other fauna have food, availability of food means less starvation, ill-health and poverty.

Anyone who doubts what I have written should consider the following: Convincing people that they have common adversary makes them united, whether the adversary is fictitious or not is irrelevant as long as people perceive it as a threat. Hitler and the Nazi Party used fake science to 'prove' that those of Jewish descent and gypsies were genetically and socially inferior and were responsible for the disintegration of the World's economy particularly Germany's, In 1930's Germany, this was treated as a universal truth, so much so that it led to the deaths of millions. By uniting a group of people against a common 'foe' the Nazis managed to gain control of Germany with a dictatorship. History is being repeated with the fictitious threat of dangerously high levels of atmospheric carbon dioxide with the ulterior motive, to gain political power.

It is a well established fact that when an opinion becomes a consensus, those who question this consensus are ostracised, belittled and/or burned at the stake, which is what happened to Giordano Bruno who was executed in 1600 for believing and teaching that the Earth and other planets orbited the Sun. Galileo was sentenced to house arrest for the remainder of his life for voicing the same opinion. This sort of scientific bigotry and intolerance were practised in the middle of the last millennium, with the Church insisting that the Earth was the centre of the Universe and punished dissenters who disagreed. In Germany during the Second World War, anyone attempting to protect Jews suffered the same fate as those they tried to protect. Religious and political persecution has been and still is rife. When this occurs in science it is even more abhorrent, because science is, or should be a logical and honest discipline. Sadly in Climatology with its scares about Earth warming it wasn't, still isn't and most probably never will be an honest discipline.

Un missionnaire du moyen âge raconte qu'il avait trouvé le point où le ciel et la Terre se touchent...

This woodcut demonstrates how the Church believed that there was a divine mechanism that maintained the relationship between the world and the sky. The artist who made this woodcut is unknown, it first appeared in Camille Flammarion's "L'atmosphère: météorologie populaire" in 1888.The translation of the caption below it reads as: A medieval missionary tells the world that he has found the point at which heaven and Earth meet.

In the third millennium AD those of us who do not believe that man is influencing the climate are labelled as 'Deniers' with all the unpleasant implications that are associated with the 'Holocaust' bring. Of course those who make a living out promulgating AGW keep using the phrase 'the science is settled' in an attempt to avoid close scrutiny of their claims. I hope after you have finished reading this book you will agree with me, that it most certainly isn't settled. Apart from house arrests and executions, very little has changed in the last 400 years with regard to scientific debate.

Studying history can be the key to understanding what the future may or may not bring as well as avoiding mistakes in the present and future which were made in the past. Earth Day was started to make people aware of the environmental damage humans were creating on Earth. At the time, they were as frightening as Climate Change is today, nearly half a century later. Here some real howlers by famous scientists from the past, starting with the first Earth Day on 22nd April 1970 this is from

https://fee.org/articles/18-spectacularly-wrong-prophecies-from-the-first-earth-day/

The Harvard biologist George Wald estimated that 'civilisation will end within 15 or 30 years unless action is taken immediately against the problems facing mankind.'

Paul Ehrlich in the Earth day issue of 'The Progressive', assured readers that between 1980 and 1989, some 4 billion people, including 65 million Americans, would perish in the 'Great Die-Off.'

In the January 1970 edition, 'Life' magazine reported, 'Scientists have solid experimental and theoretical evidence to support the following predictions: In a decade, urban dwellers will have to wear gas-masks to survive air pollution, by the mid-eighties air pollution will have reduced the amount of sunlight reaching earth by 50%.'

Ecologist Kenneth Watt declared, 'By the year 2000, if current trends continue, we will be using gasoline and oil at such a rate, that there won't be any crude oil left.'

A scientist at the National Academy of Sciences published a chart in Scientific American that looked at metal reserves and estimated that humanity would totally run out of copper shortly after 2000. Lead, zinc, tin, gold and silver reserves will be gone by 1990.

Senator Gaylord Nelson wrote in 'Look': 'Dr S. Dillon Ripley, secretary of the Smithsonian Institute, believes that in 25 years, somewhere between 75 and 80% of all animal species will be extinct.'

This has to be the best:

Kenneth Watt warned about a pending Ice Age in his speech. "The world has been getting colder for about twenty years," he said. "If current trends continue, the Earth will be about four degrees colder in 1990, but eleven degrees colder in 2000. This is about twice the cooling it would take to put us into an ice age."

This quote by Kenneth Watt was symptomatic of the then current thinking about Earth's climate; that we were heading for another Ice Age. How is it possible for 'science' to change a basic concept through 180 degrees in 20 years?

"Those in authority who dismiss out of hand those who question their right to this authority should not hold their authority in the first place."
Andrew M. Hope-Fairweather.

Chapter 2: Propaganda.

It is important to understand the mindset of those who pursue an agenda which is less than truthful.

Definition of 'propaganda': *Communicating thoughts and/or ideas in such a way to manipulate or influence the opinion and thought processes of people to ensure they support a particular cause or belief.*

The Process: Propaganda uses the following techniques, as do those who espouse climate change. It even surprised me how many of the techniques below are used by its advocates:

1) Name Calling or Stereotyping: Anyone who questions man-made climate change is called a *'denier.'*
2) Virtue Words: Ensuring people don't examine the evidence by making them think that by questioning it, they are in a minority who are wrong and are actually 'bad' people.
   'Scientific', 'sustainable' 'renewable' etc.
3) Transfer: When a symbol or a picture is used to reinforce a positive or a negative message. Both the colour and motivation; 'Green', photos of power station cooling towers emitting steam with the inference that the steam is smoke, a solitary polar bear on a small ice floe which has been photo-shopped to maximise its impact.
4) Testimonial: When someone famous or some respected organisation endorses an erroneous concept either in a positive or negative way. *+ve-BBC, MetOffice, -ve-Donald Trump, the oil and fracking industries.*
5) Ordinary People: The 'fact' that ordinary people are going to be affected adversely by a problem. *'Climate refugees', 'our children's future'.*
6) Deification: The concept or idea is not up for debate, it is a law that cannot be questioned. *'The science is settled, 97% of climate*

scientists are in agreement that mankind is changing Earth's climate', not true, this figure was pulled out of the air.
7) Bandwagon: By doing something or not doing something will impact positively or negatively on the false agenda being pursued. *'Use public transport not your car'. 'Using renewable energy will save the planet'.*
8) Artificial Dichotomy: Two sides to an argument, one of which is clearly false. By simplifying reality (reality by nature is always very complex), the prevailing incorrect view can be dominant. *'By 2070 summer temperatures in the UK will be five degrees higher than they currently are unless greenhouse gas emissions are drastically reduced'.*
9) Hot Potato: An inflammatory question whose purpose is to instil guilt in the person questioned: *'You are not still driving your 5 litre Ford Mustang are you? Just think what are doing to the climate'.*
10) Evading the question: Since there is no actual evidence of man-made climate change, the predicted temperature rises get postponed to further and further into the future. *'Why hasn't the Arctic ice cap disappeared yet? Our research continues but climate change science is still in its infancy'.* 'Still in its infancy' is a euphemism for 'we are making it up as we go along'.
11) The Lesser of all Evils: The cure is bad, but the alternative is worse. *'Ok, you cannot have electricity if the Sun doesn't shine, but the planet will die if we don't stop burning fossil fuels'.*
12) Scapegoat: Guilt by association. It transfers blame to one person or group of people without investigating the complexities of the problem, thereby deflecting scrutiny of the issue. *'Big oil and the worldwide car industries are responsible for climate change'.*
13) Confusing Cause and Effect: Which seems more likely? An increase in atmospheric carbon dioxide concentration from 0.032% to 0.04% that has caused the Earth to warm, or that we are still emerging from the last ice age that ended 11,800 years ago?

14) Cherry Picking: Publicising data that supports an idea or concept and suppressing data that doesn't.
15) False Causation and Logical Inconsistencies: *'There is more atmospheric carbon dioxide now than 200 years ago; this must be due to emissions by man, so mankind is responsible for ALL the global warming'.* This ignores the fact that carbon dioxide concentrations were much higher before humans ever existed. *'A cow eats grass. A cow is an animal, therefore all animals eat grass'*
16) Sources of Weak and Inconsistent Data: Temperature measurements adjacent to airport runways, these were fine in the 1940's when aeroplanes had propellers and flights were few and far between, but now? Satellite platinum resistance thermometers are far more accurate, but don't produce the results hoped for by those who perpetuate this particular myth.
17) Faulty Analogy: A change must always lead to an extreme. *'Tipping point, point of no return, runaway global warming, extreme weather'* This is a particular favourite of the societal engineers.
18) Misuse and confusion using statistics and dodgy graphs: Using percentages and figures without disclosing the sample size. *'97% of climate scientists are in agreement that the Earth is warming'.* That may be so, but is it because of atmospheric carbon dioxide concentration? Sample sizes are not disclosed. 9 out of 10, may be 10 like-minded friends with no scientific qualifications whatsoever, making this statement meaningless. The other favourite trick of the 'warmunists' of making the y axis of a graph start at a mid-point as opposed to zero to make the slope over the x axis steeper.
19) Fear: It cannot get much worse than the annihilation of all life on Earth.
20) Deflection: Attack the messenger rather than debate the message. Once again back to *'deniers'*.
21) Attack: Dismiss the argument as presented rather than its content. *'You would say that wouldn't you, after all you are a denier'?*

22) Pre-emptive debate: Set the tone and nature of the debate. *'We have the evidence that man-made carbon dioxide is causing the climate to warm'*. Of course they don't, but this means the discussion is undertaken on their terms and since carbon dioxide does cause warming, that is not under debate. What is never debated is by how much.
23) Diversion: If the debate is not going their way due to some embarrassing revelation, they will throw in a contentious non-sequitur, to make their opponent debate that point rather than pursuing the relevant point that was raised. *'What about the 18 year unpredicted pause in global warming, doesn't that nullify the theory'? 'Of course not, the heat passed in to the world's oceans and will eventually return to heat the atmosphere'*. Believe it or not respected 'scientits' (not a spelling mistake) believed that this actually occurred!

Chapter 3: In the Beginning.

The Birth of Our Universe and the Formation of Earth.

When our universe was created, 14 billion years ago (whether this was spontaneous, or by the actions of God is up to the opinion of each individual, the indisputable fact is that it did happen). It was unbelievably hot, much too hot for matter to exist. As the expansion of the universe continued the same amount of energy was present in an ever increasing volume which as a consequence cooled. After a few hundred thousand years the lightest atomic elements came into being, mainly hydrogen, some helium with traces of lithium and beryllium, these were the only elements present. These elements collapsed under gravity to form 'clumps' of atoms which then heated up. If the 'clumps' were dense and massive enough, their temperature rose to 14 million degrees Celsius. At this point three or more hydrogen atoms (with one proton and one electron) fuse with each other to form helium atoms (two protons, two electrons and one or more neutrons) and consequently emit a great deal of energy; this is how a star 'shines. Eventually though the hydrogen runs out and to cut a very, very long story of three billion years short, the star collapses inwards and then explodes (a supernova) producing all the heavier elements necessary to form the planets. Many of these elements are highly reactive and react with each other to form stable chemical compounds. It is thought that this matter was in orbit around the increasingly collapsing body of hydrogen which would eventually lead to the birth of our Sun.

However the exact mechanism by which our Sun and 20 million years later, our Earth and the other planets were formed is not clear. One theory is that after the Sun formed and became the centre of our solar system, a small star passed (relatively) close to the Sun disturbing the new matter and formed the planets including Earth in their various orbits, the analogy of this disturbance, would have been like ripples in a pond, with some areas of strong and others with weak gravity. The matter would have coalesced and as a result would have been extremely hot meaning that elements would have combined chemically. One of the most reactive elements is oxygen (O_2), which should not exist without being chemically combined with other elements to produce the most stable compounds possible. Like all rules there is an exception, which I will explain later. Hydrogen (H_2) will combine with oxygen to produce water, (H_2O) and carbon will combine with oxygen to form carbon dioxide (CO_2).

Like the proto-stars, gravity caused the heavier elements to coalesce and the planets were born. It was thought that a rogue planet (named Theia) struck the Earth just after its formation 4.4 billion years ago creating debris which under the pull of its collective gravity became our Moon. Iron is an abundant element on Earth and due to its relatively high mass it sank to the centre of Earth and provides our planet with a geomagnetic field. This geomagnetic field is extremely important as it deflects most of the alpha and beta cosmic rays that would otherwise strike our planet and are harmful to all life. I will describe how the Sun also performs this function later on. The geomagnetic field also deflects the Solar Wind which is a stream of highly energetic, charged particles emanating from the solar atmosphere (the Corona). Without this magnetic field the atmosphere would have been stripped away within a few hundred years making life impossible. At times of high solar activity, some of the solar wind can get through the geomagnetic field, when this occurs, the phenomenon known as the 'Aurora' (Borealis in the Northern Hemisphere and Australis in the Southern Hemisphere) can be seen. The Aurora is visible most nights near to the polar-regions and becomes less common closer to the equator.

The Aurora Borealis. Courtesy Wikipedia Commons.

The Cooling Phase.

It took Earth 500 million years to cool after its fiery formation and to allow most of the water vapour became liquid. To fully understand the timescale it is necessary to discuss geological time. Geological Time is divided into 'Eons'. The first eon is called the Hadean Eon (after the Greek God of the underworld; Hades) 4500 million years ago (mya) to 4000 mya. This is the time immediately after the Earth formed when it was extremely hot, volcanic and devoid of life. Earth's primordial atmosphere at this time was composed of the following gases: Nitrogen 78%, carbon dioxide -20%, Noble Gases (argon, neon, helium, krypton, xenon and radon) -1%. Methane, ammonia, sulphur dioxide, hydrogen sulphide and trace amounts of hydrogen comprised the remaining 1%.

Earth and the Moon as photographed from space. Courtesy of Wikipedia Commons.

Chapter 4: Life on Earth.

Building Blocks and Primitive Life.

The second eon is the Archean Eon, 4000 mya to 2500 mya, primitive life started to form by lightning during thunderstorms creating amino acids from carbon dioxide, nitrogen and water. Amino acids are the basic constituents of proteins which are essential for life.

Laboratory experiments carried out in 1952 by Miller and Urey that discharged high voltage electrodes in a flask filled with water, methane, ammonia and hydrogen, to replicate the amino acid producing gases in the primordial atmosphere together with the lightning that changed their chemistry. This experiment ran overnight, the following morning a dark brown substance was found on the inside surface of the flask, which when analysed was found to be amino acids. After Miller's death in 2007 some of the sealed vials containing the brown substance underwent analysis using equipment more sophisticated than the equipment available in 1952 and 20 more amino acids were identified.

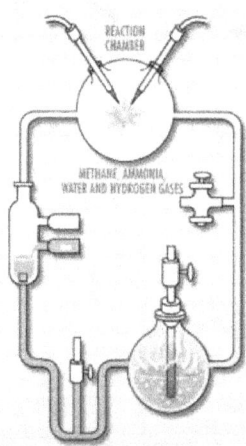

The Miller-Urey experimental apparatus. Courtesy of NIA NASA and Wikipedia Commons.

This confirmed that abiogenesis (the theory that early life can develop from inanimate matter) was how life began on Earth. There is a second theory that abiogenesis could also occur in space with radiation being the energy source to produce amino acids, some of which reached the surface of Earth. It is possible that both theories are correct and life on Earth is a mixture of evolved components of terrestrial and extra-terrestrial amino acids.

Not-So-Primitive Life.

The third eon; the Proterozoic, spans the time from the when free oxygen was first present in Earth's atmosphere to just before the proliferation of more complex life such as coral and trilobites, 2500mya to 541mya. The defining event of the Proterozoic was the presence of substantial quantities of oxygen in the Earth's atmosphere created by photosynthesis. It could not build up to any significant degree until all sources of highly chemically reactive un-oxidised elements such as sulphur and iron had reacted with the atmospheric oxygen to become oxides. Up until this time (about 2.3 billion years ago), oxygen was probably only 1% to 2% of its current level. The concentration of oxygen in the atmosphere started to increase 1.9 billion years ago, after all the iron and sulphur in contact with oxygen had all been oxidised and there was nothing left for the oxygen to chemically combine with.

Proterozoic rocks have been identified on all the continents and often are important sources of metallic ores, of iron, gold, copper, uranium and nickel. During the Proterozoic Eon, the atmosphere and oceans changed significantly. Proterozoic rocks contain many traces of primitive life-forms, the fossil remains of bacteria and blue-green algae, as well as of the first oxygen-dependent animals, the Ediacara Fauna, they were the first 'Metazoans' (animals which constituted more than one type of cell) that needed oxygen to grow. The soft-bodied Ediacara fauna were the precursors of organisms with skeletons, the appearance of which marked the end of the Proterozoic and pre-date the explosion of life that occurred later in the Phanerozoic Eon including Eurkaryotes which are life-forms that are single or multi-cellular organisms that contain organelles (these are to cells what organs are to animals). The organelles found in living cells are:

a) The Nucleus, contains the genetic material (the genome) contained within circular DNA that could split into separate sex cells, and so for the first time mixed and variable genes could be passed to younger generations.
b) Mitochondria are organelles found in large numbers in most cells, in which the biochemical processes of respiration and energy production occur.
c) Golgi bodies an organelle that processes proteins.
d) The Membrane is the outer layer that holds the cell together but allows the passage through of gases, water and nutrients.
e) Ribosomes combine amino acids together in the sequence that they are instructed to do so by RNA to make proteins.
f) Chloroplasts are found in plants and some bacteria they carry out the process of photosynthesis which will be discussed more fully later on.

By the time eukaryotes became established in the environment, atmospheric oxygen pressure had risen from low values to about 10 percent of the current atmospheric level. Megascopic eukaryotes (large multi-cellular organisms that utilised photosynthesis to metabolise) first appeared about 2.3 billion years ago and became widespread by about 1.8 billion years ago. These eukaryotes employed a form of respiration and oxidative metabolism and had a central nucleus Early organisms on Earth flourished more easily in the shallow water of continental margins. Such stable continental shelf environments (which were rare in the Archean) developed after 2.5 billion years ago, facilitating further growth of photosynthetic organisms and thus oxygen production. Evidence of the rapid rise in the oxygen includes the first appearance on continental margins of red sandstones. Their colour is caused by the coating of quartz grains with haematite (iron oxide). Other evidence is provided by the occurrence of haematite rich fossil soil beds that date from about 2.5 billion years ago. The formation of these beds is consistent with a massive rise in oxygen pressure to 0.1 atmospheres (100 millibars) between 2.2 billion and 2 billion years ago.

Photosynthesis

As these primitive life-forms continued to evolve, some became Cyanobacteria, which were capable of photosynthesis, this is the process used by plants, algae and some bacteria to harness energy from sunlight and turn it into chemical energy. Photosynthesis takes place in cells called chloroplasts.There are two types of photosynthesis: oxygenic and anoxygenic, the principles of anoxygenic and oxygenic photosynthesis are very similar, but oxygenic photosynthesis is the most common and is seen in plants, algae and Cyanobacteria. In oxygenic photosynthesis, light energy transfers electrons from water (H_2O) to carbon dioxide (CO_2), to produce carbohydrates. The carbohydrates are 'food' for the organism and the oxygen is a waste by-product. Anoxygenic photosynthesis as its name would suggest does not produce oxygen so further detail about it is irrelevant to this book. Chloroplasts are thought to have originated from primitive bacterial cells through the process of endo-symbiosis. Symbiosis is an interaction between two different organisms living in close physical association, usually to the advantage of both. Endo-symbiosis is symbiosis in which one of the symbiotic organisms lives inside the other one.

There is only one way that nature can break the strong bonds that hold one carbon atom and two oxygen atoms together to allow free oxygen to be formed and that is by photosynthesis. Photosynthesis can only occur between the temperature of 0-40 Celsius (32-104 Fahrenheit), but is at its optimum temperature range between10-20 Celsius (50-68 Fahrenheit). The current average temperature of Earth is 14.6 Celsius (58 Fahrenheit) and by implication must have been about the same when there was 500 times more carbon dioxide in the atmosphere than there is now. If carbon dioxide causes a disproportionate degree of atmospheric warming then the enzymes that metabolise carbon dioxide could not function, there would be no oxygen in the atmosphere and advanced life based on the respiration of oxygen could not exist anywhere in the universe. A human being with a body temperature raised to 40 Celsius (104 Fahrenheit), for all but the most brief of periods would prove fatal because hormones and enzymes (which regulate the human body) cannot function at these raised temperatures.

I cannot emphasise this enough, if high concentrations of atmospheric carbon dioxide cause the planet to get hotter than it is at the present time, then photosynthesis cannot occur, or can only occur in higher or lower latitudes where it is cooler. Of course the further north or south you are from the equator, the less light and heat energy there is because it strikes those areas at a much shallower angle than areas between the tropics of Cancer and Capricorn. Although the Arctic and Antarctic have 24 hours of daylight each day in their respective mid-summers, they have zero hours in their respective mid-winters. The summer mean temperature at the North Pole is 0 Celsius (32 Fahrenheit), at the South Pole it is -28.2 Celsius (-18 Fahrenheit). Winter mean temperatures at the North Pole are: -40 Celsius (-40 Fahrenheit) and at the South Pole they are -60 Celsius (-76 Fahrenheit).

In other words, if the Earth was much hotter due to more carbon dioxide being present in the atmosphere, the following would be true: Only the North and South Poles were warm enough for photosynthesis to occur BUT, the low incident light energy and its non-existence for a substantial proportion of the year would have made that impossible. However it was too hot for photosynthesis to occur anywhere else on Earth, so how did the carbon dioxide get broken down into oxygen and animate matter? The logic is inescapable, the interpretation that increased levels of atmospheric carbon dioxide cause global warming **must** be wrong.

I am going to reiterate the composition of the Earth's primordial atmosphere: nitrogen-78%, carbon dioxide-20%, Noble Gases (argon, neon, helium, krypton, xenon and radon) -1%, methane, ammonia, sulphur dioxide, hydrogen sulphide and trace amounts of hydrogen comprised the remaining 1%. Atmospheric carbon dioxide levels were once 500 times higher than they are today, yet we are here to discuss it, the only inescapable conclusion is that the current generation of doom-mongers are wrong.

The earliest evidence of life on Earth is of fossilised Cyanobacteria called Stromatolites and they are about 3.7 billion years old. Ancient as their origins are, these bacteria (which still exist today) are already biologically complex they have cell walls protecting their protein-producing RNA (ribonucleic acid). On this basis, scientists think life must have begun much earlier. In fact, there are hints of life in even more primeval rocks: 4.1 billion year old zircons in Australia contain high amounts of a form of carbon typically used in biological processes. The approximate date, give or take a few tens of millions of years may never be known.

Stromatolites found in a German Quarry courtesy of Wikipedia Commons.

Living things need RNA or DNA (deoxyribonucleic acid), complex animals, such as humans need both. The simple life-forms needed RNA only. These are two incredibly complex molecules. Their existence is a quite by chance evolution, with the odds of their existence of billions to one against. The universe is thought to contain billions of stars with billions more planets, so the odds were it had to happen somewhere Had it not happened here, life as we know it would not exist. Every human being on Earth has different RNA and DNA even if they are identical twins. Certain features change by actions and reactions in the womb, including, RNA, DNA and fingerprints! RNA is very similar to DNA, and carries out numerous important functions in each of our cells including acting as a transitional molecule between DNA and protein synthesis, and functions as an off and on switch for some genes. The

RNA hypothesis is based on the premise that RNA was the first genetic controller to evolve followed by the evolution of DNA later on. How RNA itself first arose is unknown. DNA, like RNA is a complex molecule made of repeating units of thousands of smaller molecules called 'nucleotides' that link together in very specific ways. While there are scientists who think RNA could have arisen spontaneously on primordial Earth, others say the odds of such an occurrence are astronomical meaning there is a good chance that complex life on Earth is unique to the universe.

'The appearance of a molecule this complex, given the way that chemistry functions, is incredibly improbable. It is in all probability a one-off,' said Robert Shapiro, a chemist at New York University.

A more detailed comparison on DNA and RNA can be found in Appendix 4.

Evolution on a Larger and More Diverse Scale.

The fourth eon (541mya and the one that we live in) is the Phanerozoic Eon. This is the eon where diverse and plentiful animal and plant life came into existence. The Phanerozoic Eon is divided into three **Eras**: the **Paleozoic, Mesozoic,** and **Cenozoic,** which are further subdivided into 12 periods. The Paleozoic Era features the rise of fish, amphibians and reptiles. The Mesozoic is ruled by the reptiles, and features the evolution of dinosaurs and later on mammals and birds. The Cenozoic is the time of the mammals and birds after the dinosaurs became extinct and then more recently, humans.

Eon	Era	Period		Epoch	
Phanerozoic	Cenozoic	Quaternary		Holocene	← Today
				Pleistocene	← 11.8 Ka
		Neogene		Pliocene	
				Miocene	
		Paleogene		Oligocene	
				Eocene	
				Paleocene	← 66 Ma
	Mesozoic	Cretaceous		~	
		Jurassic		~	
		Triassic		~	← 252 Ma
	Paleozoic	Permian		~	
		Carboni-ferous	Pennsylvanian	~	
			Mississippian	~	
		Devonian		~	
		Silurian		~	
		Ordovician		~	
		Cambrian		~	← 541 Ma
Proterozoic	~	~		~	← 2.5 Ga
Archean	~	~		~	← 4.0 Ga
Hadean	~	~		~	← 4.54 Ga

(Younger ↑ / Older ↓)

Image from Digital Atlas of Ancient Life

Ga is an abbreviation for billions of years ago, Ma, millions, Ka, thousands.

Evolution is change in the hereditary characteristics of biological populations over successive generations. The term 'Survival of the Fittest' is very apt to describe this process, because in its purest form those who have genetic weakness leading to mortality before having offspring means that their faulty genes are not passed on and leave the gene pool. In nature this is a very effective way for the continuance of life on Earth. Evolution has slowed to a crawl in humans simply because (quite rightly), we look after the sick and disabled, however with gene therapy becoming more and more advanced, eventually many debilitating and potentially fatal diseases will be eradicated.

Throughout the history of the Earth extinction events have taken place where up to 80% of species have completely died out. The last major extinction event took place when an asteroid or comet hit the Earth in the Yucatan Peninsula near what is present-day Mexico. The size of this projectile was estimated to be between 6.2 and 9.3 miles wide, the buried crater is called the 'Chicxulub Crater' after the name of the town currently occupying the site. The effect of this impact was felt worldwide. The extreme severity of this impact and the loss of so many species was deemed to necessitate a new Era (Cenozoic), Period (Paleogene) and Epoch (Paleocene) this took place slightly less than 66 million years ago and it is accepted that there was severe, worldwide climate disruption caused by debris in the atmosphere blocking sunlight. This event was the cause of the Cretaceous-Paleogene Extinction; 75% of plant and animal species on Earth became extinct, including all non-avian dinosaurs (those dinosaurs that had no bird characteristics, such as feathers). The crater produced by this impact is estimated to be 93 miles in diameter and 12 miles deep. Following this event, evolution went into overdrive in the Phanerozoic Eon, the age of the dinosaurs was over, the days of mammals and birds was just beginning.

Chapter 5: Complex Life.

Animal, Vegetable or Mineral?

The Carboniferous Period spans 60 million years from 359 Mya, to 299 Mya. The name Carboniferous means 'coal-bearing'. The graph below shows the concentration of carbon dioxide from 600,000,000 years ago until the present, together with average global temperatures. The Carboniferous Period was so named, because that was a time where trees grew taller and taller, because when they died and fell over fungi had yet to evolve to feed and decompose the dead wood. Dead trees did not rot they just piled up on themselves. If the new trees that replaced them did not grow tall and fast enough, then they would not have been able to photosynthesise, because the dead trees in their vicinity blocked the sunlight, so evolution favoured those species whose initial growth rates were very high.

The extremely high concentration of carbon dioxide in the atmosphere contributed greatly to this rapid growth. These trees lived in vast swamp forests, the dead trees were eventually covered by water and compressed by tectonic plates (continent sized shelves of rock that move slowly leading to earthquakes) and turned to coal seams and oil and gas deposits by heat and pressure, collectively these are called 'fossil fuels'. As can be seen in the graph below, this removed huge quantities of atmospheric carbon dioxide. I do not apologise for pointing out the obvious. As can be seen from the graph below, 530 million years ago atmospheric carbon dioxide concentration was 7000ppm, but according to the UN the increase from 1950 of 320ppm to the current 400ppm is going to make the earth too hot.

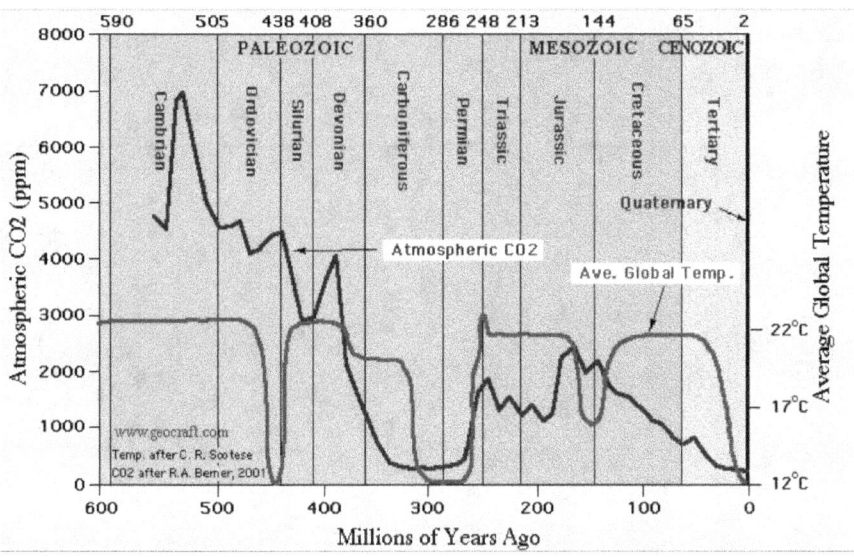

Image, courtesy of Ice Age Now.

The quantities of carbon dioxide prior to the Carboniferous were still enormous compared to present levels. Carbon dioxide levels rose again during the Triassic and Jurassic Periods, this coincided with larger, more nutritious foliage and consequently larger herbivores and carnivores (dinosaurs). The link between higher levels of carbon dioxide and a better food supply and chain could not be clearer. In addition, as the first graph demonstrates, the link between increased carbon dioxide levels and raised average global temperatures is tenuous to say the least.

The second graph (below) is an 'alarmist' depiction of what is happening. Note the starting point of carbon dioxide on the y axis at 260ppm as opposed to 0ppm as I have previously mentioned. To exaggerate the effect even further the temperature increase is accentuated by the scale in increments of 0.5 degrees with a temperature range of 2.5 degrees Fahrenheit only and the warming passing from bright blue to bright red. The year scale on the x axis of 130 years is chosen to further exaggerate the effect. This is a blatantly misleading diagram.

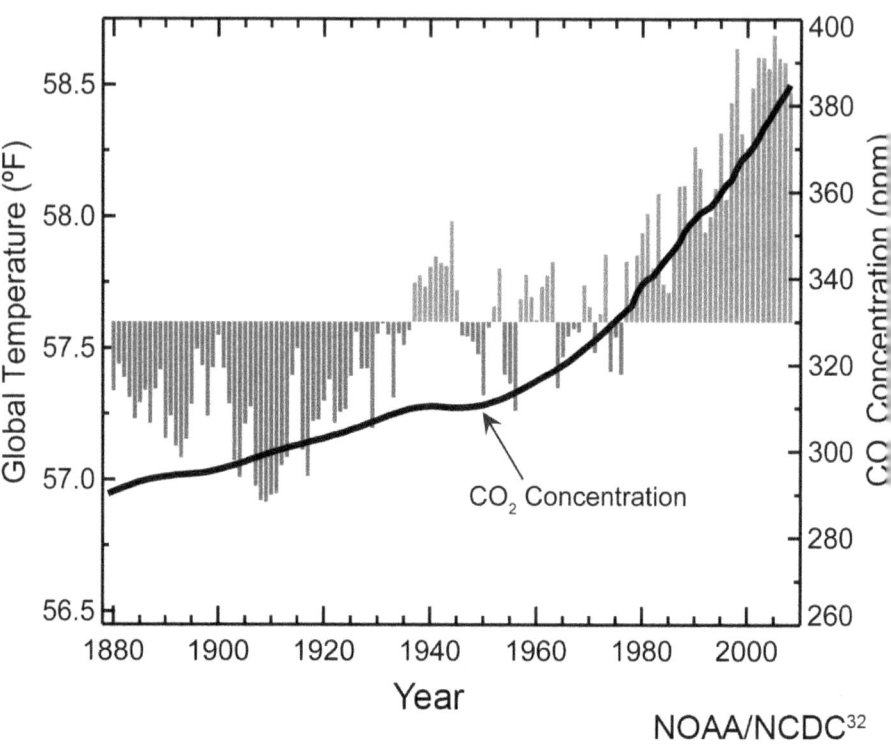

Image courtesy of Wikimedia Commons.

Crude oil was formed between 500-50 mya, and is also created through the heating and compression of organic materials over a long period of time. Most of the oil we extract today comes from the remains of prehistoric algae and zooplankton whose remains settled on the bottom of an ocean, sea or lake. Over time this organic material combined with mud and was then heated to high temperatures from the pressure created by heavy layers of sediment. This process, (diagenesis), changes the chemical composition of the material first into a waxy compound called kerogen and then, with increased heat, into a liquid (catagenesis), which is the thermal decomposition of kerogen.

The third fossil fuel is Natural Gas which is formed from the decaying remains of prehistoric plant and animal life. As with crude oil, most natural gas formation is due to the breakdown of prehistoric marine zooplankton. The zooplankton lived on a diet of phytoplankton, which, in turn rely upon the energy of the sun to produce organic materials and energy through photosynthesis. Natural gas, oil, coal and shale gas are organic materials that naturally cease any further decay. Typically, they are found at the top of crude oil reservoirs, which have been formed by the combined action of methanogenic bacteria (they produce methane while they decompose organic material) and through catagenesis. As marine sediments are buried deep within the earth, high temperatures and pressures lead to varying degrees of catagenesis.

Higher temperatures and pressures favour the formation of lighter hydrocarbons (natural gas), and so oil/gas formations that are deeper in the Earth tend to have a greater ratio of gas to crude oil. The newly formed natural gas will attempt to move to a new location. Because the Earth is completely filled by layers of solid, or at greater depths, molten rock, the gas it contains cannot exist as a self-contained liquid body but must occupy the pores that exist within the rocks. Certain kinds of rock, usually sandstone and limestone contain pores large enough and with enough connections to allow both storage and movement of fluids. Natural gas is lighter than water and rock so will move upwards until it reaches the surface, when that happens it can spontaneously burn, which is where the legend of Will O' the Wisp originated. Usually though they are trapped by an impermeable layer of rock. Methane is also produced by bacteria called 'methanogens' that decompose organic matter under anoxic conditions (no oxygen), referred to as biogenic methane. These microorganisms are also active in the intestines of most animals, released by breaking wind !

The fourth fossil fuel is Shale Gas, which as its name suggests is gas (shale oil is also present) trapped in layers of shale. Shale is a fine-grained, sedimentary rock composed of mainly ancient mud and tiny fragments of other minerals, especially quartz and calcite. It was formed by volcanic heat and geological compression. It is the most common sedimentary rock. Those deposits are much deeper than natural gas and coal and it is only with the advent of horizontal drilling technology that these deposits could be accessed in tandem with a process known as hydraulic fracturing (fracking) the process in which rock is forcibly fractured by a liquid. Many definitions describe the liquid as being pressurised, this is an elementary error since for a fluid to be pressurised its volume needs to be decreased which the definition of a liquid precludes. The process involves the injection, with a great amount of force of a 'fracking fluid' consisting primarily of water containing sand or similar material that is in suspension (to increase its viscosity) into a hole drilled into the shale deposits). This process creates cracks in the deep-rock formations through which **natural gas** and oil will flow more freely. When the **hydraulic pressure** is removed from the well, small grains of the sand hold these fractures open to allow the gas and oil to b collected.

Earth is a closed system, the carbon and oxygen atoms that were once atmospheric carbon dioxide are still in or on Earth. Some of them are fossil fuels still buried as coal, in pockets of shale and natural gas and crude oil. Burning every scrap of these will not increase atmospheric carbon dioxide to a significant degree because every living thing on our planet has the atoms of that primordial carbon dioxide in their bodies, leaves, stems, trunks etc. Even if every life form could be magically reduced to its fundamental atoms and the carbon and oxygen equally magically transformed into carbon dioxide, then our atmosphere would still not comprise 20%+ carbon dioxide because a lot of the carbon dioxide was semi-permanently* removed from its ability to re-enter the atmosphere by being formed into limestone, marble and chalk. These are what are known as sedimentary rocks, which have formed as a result of minute sea creatures with shells, dying and forming layers on the sea floor which get compressed into limestone. Add volcanic heat and the limestone becomes marble. Both these rocks are made from calcium carbonate ($CaCO_3$), the carbonate part originated from the carbon dioxide in the atmosphere.

*I say semi-permanently because it takes a temperature of 825 Celsius (1517 Fahrenheit) to achieve this, which will be the Earth's temperature in several billion years as the Sun continues to evolve and gets much hotter. I have something more to say about this later in the book.

Chapter 6: Sophisticated Living, Primitive Thinking.

As well as evolution of physical and mental characteristics, humans have undergone social evolution as well. The barbarity that has always existed in human society is thankfully confined to relatively few countries and cultures. Western society is currently not one of these cultures, where those who are less fortunate are taken care of by those who are more fortunate. However evolution has maintained the trait of putting the welfare of our offspring and our spouse above that of ourselves. The instinct to protect our own family is probably the strongest human emotion. Those who ignore this simple fact are condemned to abject failure in any form of social engineering that is attempted. A balance has to be struck in order to allow benevolence to others by minimising disadvantages to our own families. This is where the political far-Left get it wrong, the only way of defeating human instinct is by the use of force, which has characterised, without exception every one of the governments of this persuasion.

In the UK the closest we came to being governed by the far-Left was in the 1970s' when thousands of people and their families emigrated (it was called the 'brain drain') due to the punitive rates of taxation. The top rate of income tax was 83% and the top rate of tax on investment income was 98%. Dennis Healey, the Chancellor of the Exchequer famously said he would be 'taxing the rich until the pips squeak.' Anyone with any sense knew that formulating government policy based on the negative emotions of jealousy, envy and spite as opposed to positive, benevolent ones would not work. Making those who took financial risks with the consequent psychological turmoil that it entails and allowing them keep only 2% or 17% of money that they had earned was not only wrong it was also counter-productive. Taxation revenue of course fell as people avoided or evaded paying these extortionate rates either by hiring clever accountants, breaking the law, reducing productivity or by emigrating, moving their talents to other countries. Successful businesses were nationalised and as a consequence became inefficient due to lack of competition and their products and services were expensive and not fit for purpose, for the same reason. British Leyland, British Telecom (before it was de-nationalised), British Rail were by-words for incompetence. Great Britain was nicknamed 'The Sick Man of Europe.'

The next government elected in 1979 reduced the top rate of income tax on earnings first to 60% and subsequently to 40%. After each of these cuts, governmental tax receipts increased substantially. When people have more money to spend the economy grows, unemployment falls because the individual knows better than a government how to spend his/her money. Likewise governments do not run businesses very well. The Left have never learned anything from this, all people want and expect is fairness and taking money off them unjustifiably that they could spend on their own families will of course produce resentment. This is why communist regimes need force to stay in power and why prosperity eludes these countries. Who in their right mind is going to work harder to have 83% of their income taken away from them? What has this to do with the biggest scam in history? Please bear with me and all will be revealed.

The Mushroom Policy: (Kept in the Dark and fed on Sh*t!).

Those in power have used this policy to keep their citizens in check for millennia. In ancient Egypt the priests made the connection between the first sighting of the star Sirius in the autumn night sky and increased seasonal rainfall, causing the River Nile to flood the land and make it fertile. They told the people when to sow their crops based on this information, of course the observational source of this information they withheld from the peasant classes to preserve their own divine, all-knowing status.

Nothing has changed in the last few thousand years, we have alleged 'experts' whose predictions continue to defy reality, but their opinions, for some reason continue to be revered, not just in climate change either, but in every other field too. The UK Minister of Health (presumably to increase the health budget) proclaims every year that the winter influenza epidemic could claim up to 65,000 lives fortunately it is usually about one fifth of this figure. We had the Millennium Bug scare, where it was pre-supposed that all the computers in the world would fail at midnight on 31st December 1999, causing aeroplanes to fall out of the sky and massive data loss and the collapse of the NHS. Swine flu, bird flu, even global warming causing a new epidemic of Spanish Flu that killed millions in 1918 by allowing the bodies of its victims that were buried in the Arctic to thaw and release the virus back into the environment. AIDS was going to decimate the human population. Of course none of these calamities came to pass.

The MMR (Measles, Mumps and Rubella) vaccine link with autism was yet another scare over twenty years ago. On a personal note, my son was given the first of the two vaccinations, but I would not let him have the second, my reasoning being that any reaction, which we were told at the time, could lead to autism, could only happen with a second exposure. The second inoculation was to provide complete immunity the first one however would provide sufficient immunity to prevent life-threatening or life-changing complications. Many parents without medical knowledge did not know this and withheld both inoculations, so cases of measles, mumps and rubella started to rise again. I am pleased to report that my son who is now 23 years old did not contract any of those illnesses.

We have also had 'experts' telling us that a diet based on a higher intake of carbohydrates and a reduction in saturated fat was the best way to combat obesity. Predictably, obesity rates went through the roof as did Type 2 Diabetes, Rickets caused by deficiency of Vitamin D increased as did auto-immune diseases such as Multiple Sclerosis. Vitamin D is produced by ultra-violet light exposure to the skin and is a fat-soluble vitamin found in dairy produce, both sources of which were told is 'harmful' to health. The fear of skin cancer stopped children from being exposed to sunlight, so they if they were allowed to play outside, as much bare skin as possible were covered either with cloth or large amounts of sunscreen. Is it any wonder that children did not want to run around outdoors, so compounding their health problems? People are told to exercise and take up sports the result? Sport's Injury Clinics. Marathons are a significant cause of mortality, 34 deaths worldwide since the year 2000 but people are encouraged to take part in them, sometime wearing costumes raising their body temperatures to fatal levels

In the early to mid 1990s' when the concept of global warming was first discussed, I was very concerned for my two daughters aged eight and six and my newborn son. A 'Tipping Point' 'Runaway Global Warming' and the Earth being transformed into the same conditions as existed on the planet Venus were mentioned regularly on the news bulletins as were the need to cut carbon dioxide emissions to prevent this from happening. There are not many advantages to getting older but one of them is hearing all the hype associated and exaggerated within a single issue by vociferous lobbyists. Having heard similar tales before, I decided to do some research myself. I started with the planet Venus, which is considered to be Earths' 'twin'. According to the 'experts' at the time, Earth would, because of 'global warming' (as it was called at the time) become like Venus.

Photograph of Venus and Earth, comparing their actual size. Courtesy, of Wikimedia Commons.

	Earth	**Venus**
Mean Radius:	6,371.0 km	6,051.8 ± 1.0 km
Mass:	5.97 x 10^{24} kg	4.86 x 10^{24} kg
Volume:	10.83×10^{11} km³	9.28×10^{11} km³
Gravity:	9.8 m/s²	8.87 m/s²
Avg. Temperature:	14.6°C (58.3 °F)	462 °C (863.6 °F)
Temp. Variations:	±160 °C (278°F)	640 C (1184 °F)

Axial Tilt:	23.5°	2.64°
Length of Day:	24 hours	117 days
Length of Year:	365 days	224.7 days
Rotation:	Prograde	Retrograde
Water:	Yes	No
Polar Ice Caps:	Yes	No
Nitrogen:	78%	3.5%
Oxygen:	21%	0%
Argon:	0.93%	0.007%
Carbon Dioxide:	0.04%	96.5%
Air Pressure:	101.325 kPa	9,200 kPa

As seen from the data above the 'twin' comparison is not just a minor exaggeration, the two planets have very little in common that is the first myth dispelled. I have highlighted in red the proportion of carbon dioxides in their respective atmospheres and atmospheric pressures which are massively dissimilar. In the mid-nineties the scare was that the tipping point that runaway global warming would precipitate was that as more and more carbon dioxide entered the atmosphere, Earth would heat up so much that limestone ($CaCO_3$) would release carbon dioxide leaving behind calcium oxide (CaO). It took me about two minutes to find out on the internet that this was pure nonsense. The temperature required to do this (as I have mentioned earlier) is 825

Celsius (1517 Fahrenheit). The amount of carbon dioxide in the Venusian atmosphere gives it the same atmospheric pressure as would exist in one of our oceans at a depth of 2920 feet (900 metres).

The extreme atmospheric and surface temperatures have also been wrongly claimed as an example of the runaway greenhouse effect, this is also incorrect. The analogy is that if you hold a steel blade to a grinding wheel a shower of sparks is seen, place your hand in this shower and you will not get burned even though each spark has a temperature of over 1000 Celsius, because the total amount of heat energy present in each spark is negligible. The atmosphere of Venus is so dense that it can retain a great deal of heat, temperature depends upon vibration and collisions of molecules there are bound to be far more with a dense medium than a less dense one. It has nothing to do with global warming

Acidification of the Seas and Oceans.

This is another fake carbon dioxide scare, the carbon dioxide that is not causing the planet to warm because it is dissolved in water producing carbonic acid ($H_2O + CO_2 = H_2CO_3$). Acidity or alkalinity is measured in units called 'pH' (Hydrogen power). pH varies from 0 (very acidic) to 14 (very alkaline), with neutral being 7. Caustic soda is 14 and concentrated hydrochloric acid is 0, seawater is 8.2 making it alkaline, so the term 'acidification of the oceans' is nonsensical, but the term 'reducing the alkalinity of the oceans' doesn't have the same sense of forthcoming doom! The scale of pH is also logarithmic, in other words a pH of 1 is 10 times more acidic than a pH of 2. It is estimated that since pre-industrial times the seas have decreased in pH from 8.3 to 8.2, by comparison soda water (water and carbon dioxide under pressure) has a pH of about 3.5. As I have already said carbon dioxide levels were very much higher in the past and the oceans supported creatures with shells and bones, why is carbon dioxide a threat now? It clearly wasn't then, isn't now and is highly unlikely to be in the future.

Chapter 7: When Global Warming became Climate Change.

The history of man-made global warming began on the 24th of June 1988 in an article in The New York Times by Philip Shabecoff titled: 'Global Warming Has Begun'. This was based on prognostications by former US Vice-President Al Gore and the scientist, James Hansen of NASA. 'If the current pace of the build-up of these gases continues, the effect is likely to be a warming of 3 to 9 degrees Fahrenheit between now and the year 2025 to 2050. The rise in global temperature is predicted to cause sea levels to rise by one to four feet by the middle of the next century.' In his 2006 documentary *'An Inconvenient Truth'*, Al Gore claimed that sea levels could rise by up to 20 feet. He also forecast that unless the world dramatically reduced greenhouse gas emissions, Earth would hit a 'point of no return.' In his book review of Gore's documentary, James Hansen wrote: 'We have at most ten years, not ten years to decide upon action, but ten years to alter fundamentally the trajectory of global greenhouse emissions.'

What was not mentioned by these self-serving alarmists was that Earth was in an Ice Age that began 2.6 million years ago the (Pleistocene Epoch) and only ended 11,800 years ago at the start of the current Holocene Epoch the planet is still warming as it has done at the end of other ice ages. There have been at least five ice ages since Earth's formation. It is thought that ice ages are as a result of small predictable changes in the Earth's orbit and axial tilt. Geologically 11,800 years is merely a blink in the history of Earth and temperatures would be expected to rise during this time and still are.
Why the change in terminology from 'Global Warming' to 'Anthropogenic (man-made) Global Warming and then to 'Climate Change'? As I said in the Introduction the simple answer is that because the predictions made about the whole world warming were not happening, the term Anthropogenic Global Warming didn't cut any ice (sorry, I couldn't resist a terrible pun), so the term Climate Change was

introduced to be all things to all men (and women) so any climatic divergence could be blamed on carbon dioxide of the man-made variety. It is very easy to assume that the protagonists of human induced climate change are stupid, as much as it pains me to say so, they have been very clever. People will put up with expensive, unreliable energy if the alternative is much worse and what could be worse than Earth turning into a burning hell-hole? The Left benefit too, because it is the beginning of the end of their political rivals, the capitalists. Science it most certainly is not!

El Niño, La Niña (The boy, the girl)

Gilbert Walker and Jacob Bjerknes, are credited with the discovery of the El Niño Southern Oscillation phenomenon. During the early 1920s, Walker identified a periodic variation in atmospheric pressure over the Indo-Pacific which he named the 'Southern Oscillation.' In the 1960s, Bjerknes described a mechanism to explain the atmospheric features of this phenomenon over the Pacific equatorial region which he called the 'Walker Circulation'.

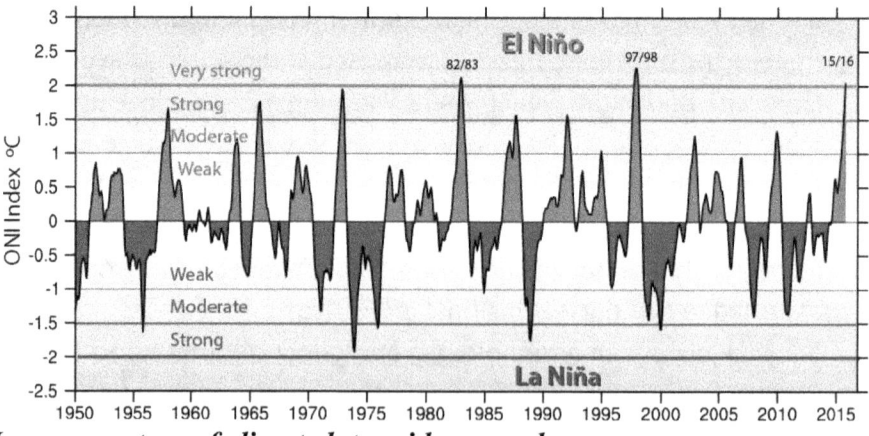
Image courtesy of climatedataguide.ecar.edu

These are weather phenomena, not climate change indicators as the 'usual suspects' would lead us to believe. El Niño is a phenomenon that causes the surface temperature of the Eastern Pacific Ocean to rise, this affects the Earth's climate on a temporary basis as does La Niña which causes cooling The last El Niño was very strong producing a great deal of warming and lasted from 2015-2016. The last La Niña was classed as 'strong' in 2010-2011 there have never been 'Very Strong' La Niñas. The Pacific Ocean is by far the largest ocean on Earth so the effects of small changes in temperature in this vast body of water are felt worldwide. It is outside the scope of this book to go into greater detail.

The Unexplained 'Pause'.

I mentioned very briefly The 'Pause' earlier, this was where average global temperatures stopped rising despite atmospheric carbon dioxide concentrations continuing to increase. This pause lasted for over 20 years and was inexplicable to the climate 'scientists' who attempted and then failed to deny that this pause existed. If carbon dioxide was solely causing the Earth's climate to warm as the alarmists claim, then this pause should not exist and most certainly not for over 20 years.

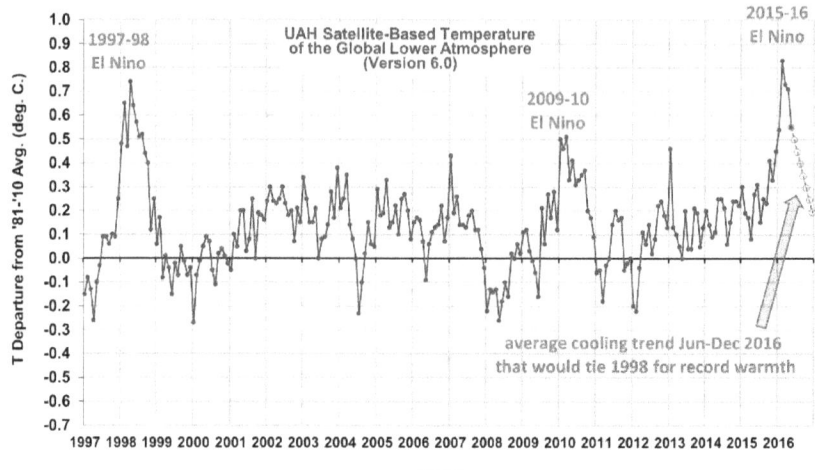

Image courtesy of gwp.com

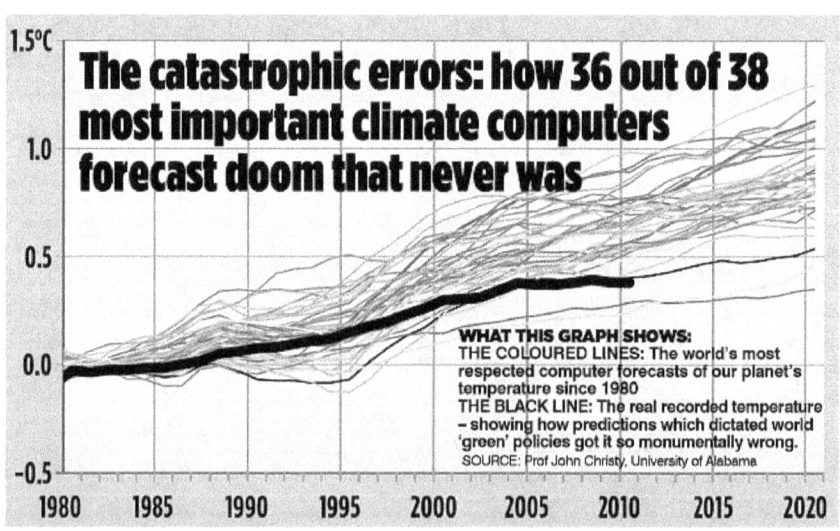

Image courtesy of The Daily Mail

The alarmism of the graph above is crystal clear, likewise the one below. As I said previously some warming of the Earth is to be expected because an ice age ended only 11,800 years ago and the graph below clearly demonstrates this. It also shows that current average temperatures are less than they were one thousand years ago, explained by solar cycles, NOT carbon dioxide. Current average global temperatures are not unprecedented as is claimed.

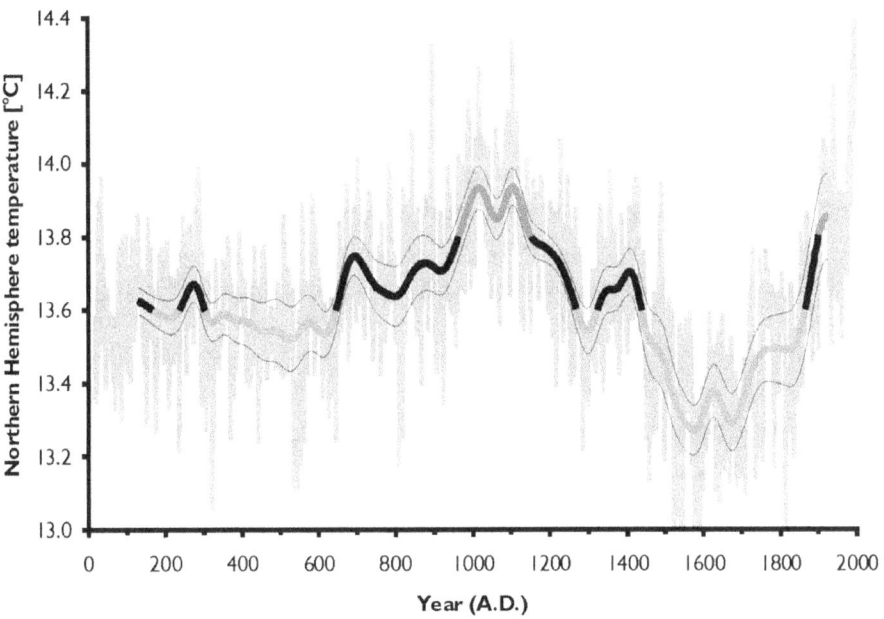

Image courtesy of Wikimedia Commons.

Chapter 8: Climate: Why Carbon Dioxide has a Minimal Influence.

I have discussed historical atmospheric carbon dioxide levels and their extremely high levels that became less and less as photosynthesis converted more and more carbon dioxide into living, dead and fossilised animal and plant tissue. I mentioned that this process cannot take place if temperatures are below the freezing point of water or above 40 Celsius and are very inefficient unless the temperature is between 10 and 20 Celsius, and it is no coincidence that the current average global temperature is 14.6 Celsius. Primordial temperatures when life first evolved had to be similar to contemporary global temperatures otherwise life would not have begun at all. Photosynthesis is such an efficient process that at one point atmospheric oxygen reached an atmospheric concentration of 35%. This level was unsustainable due to the high chemical reactivity of oxygen; making forest and grassland fires more common causing the atmospheric oxygen concentration to reduce

Temperature is based on the motion of molecules in the form of vibration when cold and faster motion when hot. I gave my children an analogy when they were at school and struggling with science. I presented them with the following thought experiment: Imagine you have a washing up bowl filled with ping-pong balls sat on a table the ping-pong balls are stationary. This demonstrates the lowest possible temperature than can exist (theoretically), it is called 'Absolute Zero' which is -273.15 Celsius (-459.67 Fahrenheit). I say theoretically, because in practice, due to quantum effects the probability of it happening is negligible. At this temperature, matter has no heat energy and so cannot vibrate (this is a gross simplification, but for the purposes of this book is more than adequate). Raising the temperature allows the molecules to vibrate, so holding the bowl and making small movements make the ping-pong balls jostle against each other, but remain more or less in their position relative to their neighbours (a solid). Increase the movement of the bowl and the ping-pong balls will make bigger movements, including some escaping the bowl (a liquid with some evaporation taking place). Increasing the motion by using yet more energy makes the ping-pong balls fly out of the bowl (boiling and turning to vapour). Most chemical processes intensify when heat is applied, but not all.

What the alarmists fail to mention when scaring the crap out of us all is that the effect of atmospheric carbon dioxide concentration on global temperature is logarithmic NOT linear (in short doubling carbon dioxide concentration does not double the amount of warming it produces). This mechanism is as follows: Our Sun emits a number of different wavelengths of radiation one of which is infra-red radiation which heats Earth and the other planets, moons and asteroids in our solar system. Some of this infra-red radiation is radiated back into space, by these bodies, especially during their respective night-times.

This radiation is called 'Infra-Red, Black Body Radiation'. The amount of radiation emitted is dependent upon the temperature. Greenhouse gas molecules absorb the photons of the infra-red radiation preventing them from radiating out to space and instead the atmosphere warms. As a basic fact, this suggests that significant warming will occur. However this simplistic explanation does not work for the following reasons:

1) The absorption of infra-red radiation by carbon dioxide molecules only occurs with three photonic wavelengths, 2.7, 4.3 and 15 micrometres (μM). Infra-red radiation is defined as any radiation with a wavelength of between 0.7 micrometres (0.0007 millimetres) and 1 millimetre retain the heat causing the greenhouse effect, which as I have previously stated I do not challenge. However, what I do challenge is that this effect involves more than 0.063% of atmospheric carbon dioxide, in other words 99.937% of atmospheric carbon dioxide molecules do not trap heat.
2) A further cause and exaggeration of this logarithmic effect is as follows: Those photons of the 'correct wavelength' will be absorbed by a carbon dioxide molecule as they travel at the

speed of light back into space, this can only happen once. In other words carbon dioxide molecules closest to the Earth's surface act as 'shields' to any carbon dioxide molecules further along the same trajectory, so any carbon dioxide molecules above them are an irrelevance as far as trapping heat goes. The 'altitude' where carbon dioxide molecules become an irrelevance is 10 metres (32.5 feet), double carbon dioxide concentration to 800 ppm and it becomes five metres (16.25 feet). In our atmosphere, the lowest layer of our atmosphere (the Troposphere) extends from sea level to 12,000 km (39,000 feet) in the remaining 11,99.995 km (38,983.75 feet) of the Troposphere any carbon dioxide present absorbs no photons whatsoever. When carbon dioxide was at pre-industrial levels of 280 ppm all the absorption occurred at a height of 12.5 metres (20.31 feet). I would not describe this as 'catastrophic' by any stretch of the imagination! Before photosynthesis had metabolised carbon dioxide into plant food and oxygen and its atmospheric concentration was 20% (200,000ppm) all the photons would have been absorbed at a height of 19cm (7.5 inches). If humans had been around then, they would have had warm feet and ankles!

3) In some cases the hotter carbon dioxide molecule releases another infra-red photon which has to have less energy and therefore a different wavelength so it is not absorbed by carbon dioxide molecules higher up in the atmosphere either.

As can be seen from the graph below the greatest amount of warming occurs when carbon dioxide is 20ppm, increasing its concentration has a negligible effect.

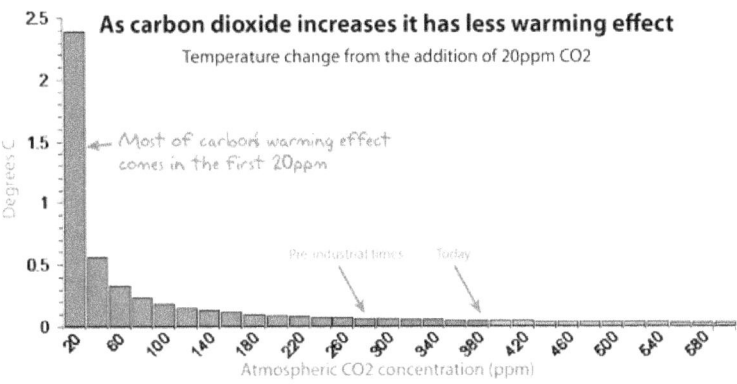

Image Courtesy from Iceagenow.

During the Ice Ages, atmospheric concentration of carbon dioxide was 180ppm. 30ppm less and our world would have become lifeless, because if carbon dioxide is less than 150ppm photosynthesis is impossible, this would have ended all life on our planet because photosynthesised material is at the base of the food chain for all living creatures. Historically its current levels are extremely low and food production and its associated benefits would dramatically increase if it was higher.

I have also checked up on isotopes of carbon and oxygen (atoms with more or less neutrons) to ensure that they could not cause significant warming. Carbon has 15 isotopes with between 8 and 22 neutrons 12 of which don't occur naturally. Carbon 12 (98.9%) is the commonest and is stable as is Carbon 13 (1.1%). Radioactive Carbon 14 is the third natural one found with a half-life of 5730 years and is only found in small quantities and is the basis for carbon dating archaeological relics. Oxygen has three naturally occurring isotopes; Oxygen 16, 17 and 18 with 16 being the most common at 99.762%. It is clear that these isotopes are so rare that they cannot increase warming by any significant amount.

Chapter 9: Science or Politics?

I wrote this book with the intention of warning that our societies were going to change for the worst, based upon a lie of the highest magnitude. My initial intention was to write about the science, or rather lack of it, behind some of the doom-laden predictions for our planet and all of its inhabitants. Politics is undoubtedly the main driver behind Global Warming, Anthropogenic Global Warming, Climate Change (call it what you will?). The purpose of the emotive evolution of these terms is purely political there is no credible science behind it whatsoever.

The United Nations.

After the end of the First World War an organisation was formed to prevent such a terrifying event ever happening again, this organisation was called: 'The League of Nations' and was formally created in 1920 as a result of the 1918 Treaty of Versailles which ended the First World War. Its remit was to prevent another world war it clearly failed to do this with the outbreak of the Second World War in 1939 and as a consequence was disbanded. The United Nations was the organisation that superseded it as an instrument of the political Left who want to take control of world governance. The UN has evolved into a self-satisfying, Left wing organisation and like all political extremist organisations they collectively think that they occupy the moral high ground and that the means always justifies the end. The term United Nations was first suggested by US President Franklin D. Roosevelt in the Declaration by United Nations on 1st January 1942 when 26 representatives of the allied powers promised to continue their alliance to fight the Axis Powers led by Germany. It came fully into being on the 24th October 1945 and originally had 51 countries as members (currently this has risen to 193). Its headquarters are in the UN Building, Manhattan, New York City. The UN has five principal sections.

The General Assembly: This is the main policymaking and representative organ of the UN. It makes decisions on important questions, such as those on peace and security, admission of new members and budgetary matters require a two-thirds majority. Decisions on other questions are by simple majority.

The Security Council: This makes decisions regarding certain resolutions for peace and security.

The Economic and Social Council (ECOSOC): This promotes international economic and social co-operation and development.

The Secretariat: This provides studies, information, and other clerical facilities needed by the UN.

The International Court of Justice: The primary judicial court.

The UN also has various System Agencies, including the World Bank, the World Health Organisation, The World Food Programme, United Nations Educational, Scientific and Cultural Organization (UNESCO) and United Nations International Children's Emergency Fund (UNICEF).

Its powers are much greater than that of the League of Nations and all of its different component parts give it the status of a very powerful government complete with its own army.

The UNs' Left leaning agenda is not particularly well hidden. It regards worldwide Socialism with a World Government (led by the UN of course) as desirable and wants equality for everyone on Earth. The problem is that rather than making poorer countries richer they want to make richer countries poorer. The European Union was the model for this, but with Brexit this is now in doubt, plus there are other populist forces gaining the upper hand including President Trump, hence the UN's renewed urgency to move 'forward'. Some people may think this is a good thing, but it is also very, very dangerous. The status quo with the major world economies being capitalist has resulted in the numbers of people on Earth living in absolute poverty (those earning less than $1 a day, updated in 2015 to $1.25 a day) being drastically reduced. The reduction has been dramatic, from 84% in 1820 to 24% (Bourguigon and Morrison, 1995). A more contemporary study by the World Bank indicates that the number of the World's population earning less than $1.90 has reduced from 44% in 1980 to 9.6% in 2015. These results are even more impressive because there were seven times more humans on Earth in 2015 than there were in 1820. Of course this is still too high, we have a long way to go, but fake science impeding this progress is most certainly not going help achieve it. Much of this progress has been made by the availability of cheap and plentiful energy which will cease if those who pursue these ill-thought out policies succeed.

Communist countries such as Russia have had to embrace capitalism to become more prosperous. There are only five remaining truly Communist countries and most of those allow private ownership, the biggest one China has some capitalist features, which allow it to prosper which it didn't when Mao Tse Tung was in charge. Communism is a theory or system of social organisation in which all property is owned by the community (the State) and each person contributes and receives according to their ability and needs. This sounds good in theory but in practice it has failed time after time, because it is a system that is contrary to human nature, despite its sound logic at face value.

Communist Countries Present and Past.

China, Cuba, Laos, North Korea, and Vietnam are still communist.

Armenia, Azerbaijan, Belarus, Estonia, Georgia, Kazakhstan, Kyrgyzstan, Latvia, Lithuania, Moldova, Russia, Tajikistan, Turkmenistan, Ukraine, and Uzbekistan are independent countries that were under the control of the Soviet Union. The Russian Federation still exerts some control over these countries despite them having independent governments.

Afghanistan, Cambodia, Mongolia, and Yemen are Asian former communist countries.

Bulgaria, Czech Republic, Germany (East), Hungary, Poland, Romania, Slovakia, are all totally independent countries. Germany was split into two separate countries after the Second World War. West Germany was controlled by France, UK and USA, East Germany by Russia. Berlin the capital was in East Germany and was similarly divided in two. Germany was re-unified with the fall of the Berlin Wall in 1989.

Balkan countries which were ruled by communists are: Albania, Bosnia and Herzegovina, Bulgaria, Croatia, Rep. of Macedonia, Montenegro, Serbia, and Slovenia.

Formerly communist countries in Africa are: Angola, Benin, Dem Rep. of Congo, Ethiopia, Somalia, Eritrea, and Mozambique.

These countries learned the hard way that Left wing policies don't bring prosperity, the UN is clearly swimming against the tide, but as I have already said and it is worth repeating, the only function of the global warming scare was to impoverish the civilised world, by making our computer-run societies rely on intermittent and expensive energy. I have, when discussing this been told I am a 'conspiracy theorist' a 'tin hat wearer' and a few other things that would more likely than not, prevent this book from being published. The evidence that there is no science but plenty of politics is this webpage:

https://motherboard.vice.com/en_us/article/43pek3/scientists-warn-the-un-of-capitalisms-imminent-demise

The link below is the direct link to UN Agenda 2030 the implementation of which is intended to impoverish the West. The association between cause and effect could not be any stronger

https://sustainabledevelopment.un.org/post2015/transformingourworld

This clearly shows that there is nothing scientific about man-made climate change it is purely political. The BBC radio news has been banging on about extinction and climate change. There is no concrete evidence proffered by the World Wildlife Fund (WWF) or the Intergovernmental Panel on Climate Change (IPCC), they make these glib statements and everyone unquestionably accepts them. The Mushroom Policy is sadly still alive and well, those in charge must be right? No, they are liars or incompetent, either way they should be punished for crimes against humanity.

They state that the rate of climate change is increasing. The climate is not changing any more or more rapidly than it normally does. Species have always become extinct, that is not occurring any differently either. Inequality is not rising it is falling as evidenced by Gay Rights, LBGT groups and increasing tolerance of religious and ethnic minority groups. Unemployment, slower economic growth and rising inequality will be the result if energy becomes expensive and intermittent which it will be with a 100% certainty by reliance on renewable energy.

They also claim on the motherboard website that capitalism cannot survive because of increasing energy costs caused by the 'necessity' of switching to renewable energy sources. Well of course it can't survive under these conditions, which is the sole intention of the global warming scare! The doom-mongers have created a self-fulfilling prophecy based on their enforced solution to a non-existent problem. They want to destroy a system, which I agree is not perfect, but has achieved results that the Left most certainly cannot remotely achieve.

A comparison between North Korea with South Korea is a good way to see the difference between far Left and moderate Right. They are two countries similar in land area they speak the same language they were both founded at the same time, (they both came into existence at the end of the Second World War when Japan surrendered and as part of the conditions lost its control of the Korean Peninsula which it had occupied since 1905). The Soviet Union gained control of the North and the USA of the South, which is why South Korea is democratic and North Korea isn't (North Korea calls itself 'The Democratic People's Republic of Korea' which is either irony or the result of a warped sense of humour).

	South Korea	**North Korea**
Population:	51,181,299	25,248,140
Birth Rate:	8.3 /1000	14.6/1000
Death Rate:	6/1000	9.3/1000
Infant Mortality Rate:	3/1000	22.1/1000
Number of Years in Education:	17	11
Life Expectancy in Years:	82.5	70.7
Maternal Mortality Rate (Live Births per 1000) :	11	82
GDP:	$2.027 Trillion	$40 Billion
GDP per Capita:	$39,400	$1800

South Korea produces 50,675,000 times more wealth than North Korea, if the figure is adjusted to take into account the differing population numbers South Korea is over 25 million times more productive than North Korea. Where would **you** rather live and which country is more able to deal with poverty, access to free or affordable healthcare, social security and all the other benefits that wealth brings? This is the reality of the Left a Socialist Utopia has never existed and never will. It is a myth that wealth can be redistributed under communism and to a lesser extent under socialism, you cannot redistribute something that does not exist because no-one in their right mind is going to take a financial risk if there isn't any significant benefit to their family or themselves. Where is the incentive to work if the fruits of your labours are taken away from you and given to unknown and possibly undeserving strangers?

Chapter 10: The Rationale versus the Rational.

The next question is why the deceit, fraud and lies?

Greed and grant money which both flow like water for scientists studying the 'problem' are part of the problem. So far it has succeeded because people are frightened to question something that is considered to be orthodox thinking, especially if they are then compared to Nazis. Conspiracy theories and fake news are so common these days that people do not want to be tarred with the 'lunatic' brush. To be completely frank, I was the same. "The am I missing something here?" thought kept popping up in my mind every time I thought about writing this book, when I actually was writing it and again later on when I thought about publishing it.

I have questioned the 'science' in the first seven chapters of this book, now I am questioning the logic of the actions, or rather the lack of them, taken to negate the alleged effects of man-made carbon dioxide and prove the political reality.

1) Renewable energy sources rely on wind, Sun and tides. 21st century technology is not capable of functioning using medieval methods. If the wind doesn't blow, the Sun doesn't shine and the phase of the Moon is 'wrong' we don't have any power and our computer driven societies grind to a halt and as a deliberate consequence, so do capitalist economies. Where is the logic in making the prosperous countries of the world that send financial aid to those countries who are poor and deprived unable to do so because that option has been taken away from them due a non-functioning economy?

2) Wind turbines are built in remote areas that no-one through choice would want to go to. Each one needs 800 tons of concrete in their foundations and specially strengthened roads

that can bear the weight of the turbines and the vehicles building and maintaining them. They can only generate electricity when the wind is blowing between 30 and 50mph. If there is no wind the turbine blades have to keep turning or the weight of the blades deforms the bearings causing them to catch fire due to the resultant frictional heat produced. If the wind speed is too high then the turbines have to be braked, again to stop the bearings overheating. Both these processes use power from the National Grid because the generators become motors (electric generators and motors are physically identical). They cannot be built close to where people live because they negatively affect human health and that of other creatures too, but for those who say we need them (and also massively profit from them), their impact on other living things is an irrelevance. Worse still they actually kill these creatures not by the blades striking them, but by the sudden dramatic change in air pressure the blades produce causing the bird's and bat's lungs to implode, which I would imagine is a particularly horrific death. A report today says that this is actually affecting the balance of nature because the eagles, falcons, hawks, owls and the other apex predators are being disproportionately killed allowing rodent populations to increase. This of course, like all the other negativity associated with renewable energy has all been glossed over! The claimed longevity of wind turbines (especially the offshore ones) is questionable and the amount of carbon dioxide emissions they save is in all probability very similar to the carbon dioxide their manufacture, deployment and maintenance that they, their roads (that ordinary traffic never uses) and the pylons and cables that they need, produce in the first place.

3) Modifying power stations to burn wood as opposed to coal seems like a good idea; trees are after all renewable! The implementation of this however isn't, the wood comes from the United States, all 7,500,000 tons of it to the UK per annum. The mature and atmospheric carbon dioxide-removing trees are

felled, dried in ovens, pulverised and then reconstituted into pellets. These pellets are transported by road and rail to ports on the US East coast and then shipped 3500 miles across the Atlantic. Coal would provide 30% more energy per ton than wood does and there is plenty of it in Europe. One of the most recent conversions to wood generated electrical power is Drax Power Station in the middle of the Yorkshire coalfield where much wiser people situated it. This coal could still be mined, albeit not as easily as it was in the past, but it would still be cheaper and kinder to the environment than the present alternative, which also involved expensive modifications to change it from coal to wood burning. The American felled trees are re-planted but take 30 years to grow back to the size they were when they were felled and made into pellets. As saplings their consumption of atmospheric carbon dioxide is negligible compared to a mature tree. Stupidity or insanity, take your pick?

4) For cheap, reliable and safe electric power why haven't nuclear power stations been commissioned? I accept that there are some (minimal) risks with conventional uranium reactors, but not with thorium reactors which cannot meltdown and do not require boron rods to control nuclear fission. The technology for these reactors has been available for 70 years. They were never commissioned before because collectively the world's military opposed their use their reason being that thorium, unlike uranium, does not produce fissile material necessary for the manufacture of nuclear weapons. Thorium is abundant and these reactors can be made small enough to supply a village or large enough to supply a major city. The carbon dioxide produced by their construction, maintenance, eventual decommissioning and the mining and purifying of thorium will be more than negated by their lifetime generation of electricity compared with the generation by the equivalent in fossil fuels.

5) Given that most people believe (erroneously) that the fate of all life on the planet is at stake, surely all but essential flights should be grounded, recreational cruises banned, petrol rationed and a lot of other restrictions put into force. They haven't been; why? Why can't the meetings to discuss climate change like the ones in Paris and currently Poland be carried out online rather than the fraudsters flying all over the world to discuss it? Do what I say, not what I do springs to mind.

6) It is ironic (or more likely, moronic) that our Left leaning, climate change believing BBC refused to renew their contract with the equally Left leaning, climate change believing Met Office in February 2018 and have now contracted Meteo Group for their weather forecasts; the reason? The Met Office forecasts were inaccurate. This was despite the fact that the Met Office had a new powerful super computer installed in December 2016. The problem was that the Met Office, believing their own hype, incorporated the fictitious increasing global temperatures as part of the computer programming. Crap in = crap out!

Chapter 11: The Devil In(car)nate?

The motor car has been the symbol of freedom and independence for over a century. In 1895 there were three types of car; internal combustion engine vehicles, steam powered vehicles and electric motor vehicles. Steam rapidly fell out of favour due to the 30 minutes it took to light the fire that heated the water tank and build up a head of steam. They were not much use as emergency vehicles either! That left internal combustion and electric vehicles. The latter were particularly favoured by women, who wisely tend to be more risk averse than men because they did not need to be hand cranked with a starting handle which required a degree of strength and quick reflexes to avoid a fractured thumb or wrist if the starting handle kicked back. Their lead-acid batteries could be re-charged overnight and had a range of about 100 miles. Internal combustion engines did and still have the advantage of easy and much quicker refuelling times and after the invention of the first patented electric starter motor in 1903, they quickly found favour with women too. The problem with electric vehicles is that the batteries are very heavy and their contents on contact with human skin are highly destructive. This was just as true then as it is today. It is claimed that electric vehicles are better for the planet, this simply is not true, another deceit to add to the growing list.

Rare metals (also called 'Rare Earths') are necessary for construction of the batteries (that have a finite life and are difficult to recycle) and these rare earths only exist in tiny quantities and usually in inconvenient places; so you have to move a lot of earth to get just a little bit of the rare variety. In China at the Jiangxi rare-earth mine the workers dig eight-foot holes and then pour ammonium sulphate into them to dissolve the sandy clay. They then drag out bags of this clay solution and pass it through several acid baths. What is left is baked in a kiln, evaporating the liquid and leaving behind the rare earths. At this mine, those rare earths amount to 0.2 percent of what gets pulled out of the ground. The other 99.8 percent, now contaminated with toxic chemicals is dumped to contaminate the environment. In other words every 1000kg of soil processed yields 2kg of rare earths.

In every stage of the process, mining has emissions that are hidden. Jiangxi has relatively few because clay is relatively easy to process, but many mines need rock-crushing equipment with consequent astronomical energy consumption, as well as coal-fired furnaces for the last of the baking stages. Those emit a lot of carbon dioxide in the process of refining a material destined for an allegedly zero-emissions car. In fact, manufacturing an electric vehicle generates more carbon emissions than building a conventional car, mostly because of the construction of its battery. The battery will of course consume more power if the air-conditioning is used and in winter the battery will lose power quickly because of the heater which in a conventional car uses the by-product of heat produced by the engine, which is negligible from an electric motor. In addition, a battery relies on a chemical reaction which becomes much more inefficient at low temperatures and over time the battery becomes more inefficient and holds less charge.

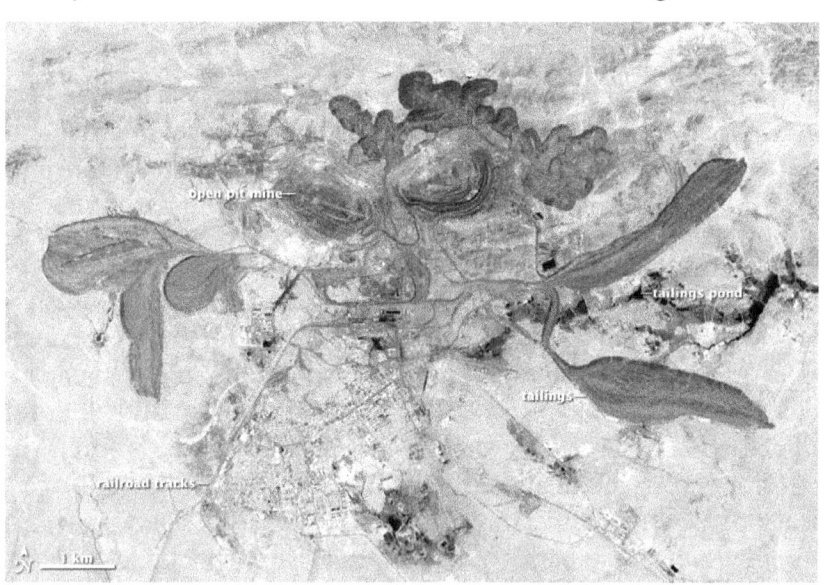

Rare Earth Mine Bayan Obo, China. Courtesy of Wikimedia Commons.

There is also the inescapable truth regarding the Law of Conservation of Energy, one consequence is that when energy is transformed into another form, there are always losses. A car with an internal combustion engine transforms potential energy in the fuel into heat, then into motion, firstly to the pistons in the cylinders and then to the driven wheels. If the wind is blowing and the Sun is shining with a full or new Moon, renewable energy sources can provide electricity. If not, potential energy from coal or wood is changed to heat energy, then into energy of motion (turbines in a power station), then to chemical energy (charging the batteries) and then to energy of motion of the electric motor and then to the driven wheels. There are more energy losses from electric vehicles than from conventional ones especially in winter when the range and efficiency of them is drastically reduced.

Burning one tonne of oil produces 11.63MWH (Mega-watt hours). In the UK in 2001 (the last year I have information from), road transport used a 42 million tonnes equivalent. This equates to 488,460,000 megawatt hours or 488,460 Gigawatt Hours. In 2017 total production of electricity by wind produced about 10% of the power needed to supply power to all road vehicles, with nothing left, to provide electricity for homes, factories, hospitals and other consumers. If petrol and diesel were outlawed today our economy would grind to a halt with the consequence that the less-able would suffer the most. This should cause alarm amongst the Leftist 'intelligentsia', but it won't because it never has in the past. The Left regard their principles as their primary objective as opposed to people's needs.

Chapter 12: More Drivel from the 'Experts.'

Failed predictions are always good material to discredit a falsehood:

In 1986, NASA scientist James Hansen testified before Congress that global temperatures should be nearly two degrees higher in 20 years. He also said that in 20 years, the area below his New York City office would be completely changed, most notably that the West Side Highway adjacent to the Hudson River will be under water.

Carl Sagan predicted in 1990 that the planet could face an ecological and agricultural catastrophe within the next decade.

Also in 1990, Dr. Michael Oppenheimer wrote: By 1995, the greenhouse effect would be desolating the heartlands of North America and Eurasia with horrific droughts, causing crop failures and food riots. By 1996 The Platte River of Nebraska would not exist, while a continent wide black blizzard of dry soil will stop traffic on highways, strip paint from houses and cause computers to shut down. Americans will migrate illegally to Mexico to escape the disaster befalling their country.

Dr. Hansen and Peter Wadhams, the head of the Polar Ocean Physics Group at the University of Cambridge, stated that the Arctic is likely to become ice-free by 2015.

Al Gore predicted in 2007, 2008 and 2009 that the scientific consensus believed that the North Pole would be 'ice free by 2013.'

In a 'scientific' report in 2003 the IPCC warned, that within the next decade, parts of California would be flooded, some of the Netherlands would be uninhabitable, and a rise in hurricanes, tsunamis, and tornadoes would provoke wars across the world as people fought for increasingly scarce resources.

'Scientists' with the United Nations Environment Programme warned in 2005 that man-made global warming would create more than 50 million 'climate refugees by 2010.

Contradictions are also helpful in deciding if a problem is valid, imagined or purely propaganda.

Para-phrased from the website: climatechangepredictions.org (Having it Both Ways)

Trees Have Less Colour
Scientists at the University of New Hampshire have said that climate change will make the leaves of trees in New England and New York less colourful over the next century. New York Times, 16 Oct 2005

Trees More Colourful
The Tree Council said that the lack of moisture in autumn caused by climate change means that a different pigment is produced called anthocyanin a policy adviser at the UK Woodland Trust.
Climate change models for the UK suggest we are going to have hotter and drier summers, which will produce the colours normally seen in a New England fall.
The Guardian 18 Nov 2004

Winters Warmer
Climate change means that we expect winters to be warmer and wetter said a Met Office meteorologist.
BBC News, 27 Feb 2007

Winters Colder
Britain's winters will get colder because of melting Arctic ice, the Met Office chief scientist; Julia Slingo said. Climate change was moving the UK towards freezing and drier winter weather and called publicly for the first time for an urgent investigation.
The Sun (UK), 11 Apr 2013

Snowdon Going Downhill
Experts from the Bangor University suggest that a white Christmas on Snowdon, the tallest mountain in England and Wales may one day be

just a memory. The figures indicated that this winter, Snowdon is likely to have less snow than in any of the last 10 years.
BBC News 20 Dec 2004

Snowdon Going Uphill
Snowdon Mountain Railway will be shut over the Easter weekend after it was hit by 30ft (9.1m) snow drifts.
BBC News North West Wales, 28 Mar 2013

On a more serious note, sadly for science this probably sums it all up: From the website: climatechangepredictions.org (In Their Own Words)

A lot of climate change research is based on opinion. Climate models sometimes disagree on the signs of future changes. Only sensational exaggeration makes the kind of story that will get politicians, and public attention.
Monika Kopacz, NOAA Program Manager 2009 April 12 2009

No matter if the science of global warming is all wrong there are still environmental benefits to having global warming policies. Climate change provides the greatest opportunity to ensure justice and equality in the world.
Christine Stewart, then Canadian Minister of the Environment Calgary Herald, 1998

As scientists we are ethically obliged to uphold the scientific method, in effect promising to tell the truth.
However, we are human beings as well as well as scientists and like most people we'd like to make the world a better place. To do this we need to get some broad-based support, to capture the public's imagination by getting a lot of media coverage. Therefore we have to make up scary scenarios, make simplified, dramatic statements and not mention any doubts we might have.

Dr. Stephen Schneider, former IPCC Coordinating Lead Author, APS Online, Aug./Sep. 1996

Two other good websites are:

Watts Up With That (WUWT).

Drunken fish? You've got them!

https://wattsupwiththat.com/2012/01/17/co2-increases-to-make-drunken-clownfish/

Deaf fish? You've got them too!

https://wattsupwiththat.com/2011/06/04/co2-deafens-nemo-or-how-many-ichthyologists-can-you-fit-in-that-car/

Climate Change affecting other civilisations?

https://phys.org/news/2014-06-climate-contact-alien-civilisations.html#jCp

Climate Change increasing the risk of an asteroid strike? Here it is!

https://wattsupwiththat.com/2012/12/22/stupidest-global-warmingclimate-change-story-of-the-past-decade/?cn-reloaded=1

Climate Change and the ability of sharks to hunt. Here is the link.

https://wattsupwiththat.com/2015/11/12/climate-science-jumps-the-shark-sharks-hunting-ability-destroyed-due-to-higher-co2/

NOT A LOT OF PEOPLE KNOW THAT

On a more serious note!

https://notalotofpeopleknowthat.wordpress.com/2018/11/26/goves-climate-nonsense/#more-36429

https://notalotofpeopleknowthat.wordpress.com/2018/11/25/we-cant-blame-wildfires-on-global-warming-booker/

Here are some predictions from the University of East Anglia, Department of Doom all the statements were made in the early to mid 2000's.

Within a few years snowfall in winter will become a rare and exciting event. Children just aren't going to know what snow looks and feels like.

Climate Change could dramatically change British tourism in the next 20 years, with European tourists holidaying in the UK to escape unbearably hot continental summers.

Weather changes may provide benefits for northern seaside towns such as Blackpool and put more strain on roads in southern coastal resorts as countries like Spain become too hot for holidaymakers, a study in the Journal of Sustainable Tourism said.

The likelihood is that in the Mediterranean countries, summers may be too hot for tourists after 2020.

Climate change will affect the Scottish ski industry. It is very vulnerable to climate change; the resorts have always been dependent on snowy winters and as the rate of climate change increases, it is hard to see a long-term future. It was reported in 'The Scotsman' on 13/06/2018 that in the winter of 2017-2018 the Scottish ski industry had its best season in five years.

I am writing this in Marbella, Spain, where our reservoirs are full and the 20 year old forecast of desertification that we should be suffering from here and now, is not happening. Cloud-free days total 342 out of 365. By September of this year (2018), there had been many more cloudy days than 23.

Chapter 12: Global Cooling – potentially the real problem.

The absolute worst case scenario is that the UN succeeds with their course of action and as a result we have expensive and intermittent power and the climate cools. Humanity is stuffed! Could this happen? There is a possibility that it could and this is how this would occur.

The Sun is very complex and not very well understood. Solar Cycles were discovered in 1843 by Samuel Heinrich Schwabe. After 17 years of observations he noticed a periodic variation in the average number of sunspots. Sunspots are basically magnetic storms on the Sun. Rudolf Wolf compiled and studied these and other observations, reconstructing the cycle back to 1745. Eventually he pushed these reconstructions back to the earliest observations of sunspots by Galileo and other astronomers in the early seventeenth century following the invention of the telescope. Using Wolf's numbering scheme, the 1755–1766 cycle is Cycle Number 1. Wolf also created a sunspot number index called the Wolf index, this is still used today. We are currently approaching Solar Cycle 25

Between 1645 and 1715, there were very few sunspots, this period is called 'The Maunder Minimum, after Edward Walter Maunder, who researched this event which was first noted by Gustav Spörer.

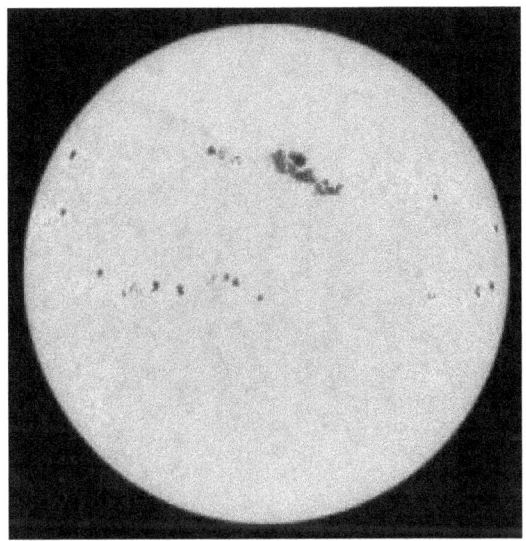

Sun and sunspots courtesy Wikipedia Commons

Simplifying the physics as much as I can, sunspots are very important with regard to the Earth's climate. Sunspots envelop the Earth with their magnetic field, the more sunspots there are, the stronger the field. This magnetic field deflects two kinds of cosmic rays (radiation emitted by violent stellar events in the universe). These are alpha particles (positively charged particles consisting of two protons and two neutrons) and beta particles (high energy electrons), If these particles enter our upper atmosphere, they 'seed' cloud formation. The classic experiment demonstrating this is the 'Wilson Cloud Chamber'. The third type of cosmic rays, gamma rays is an electromagnetic radiation, not particles, so does not seed cloud formation.

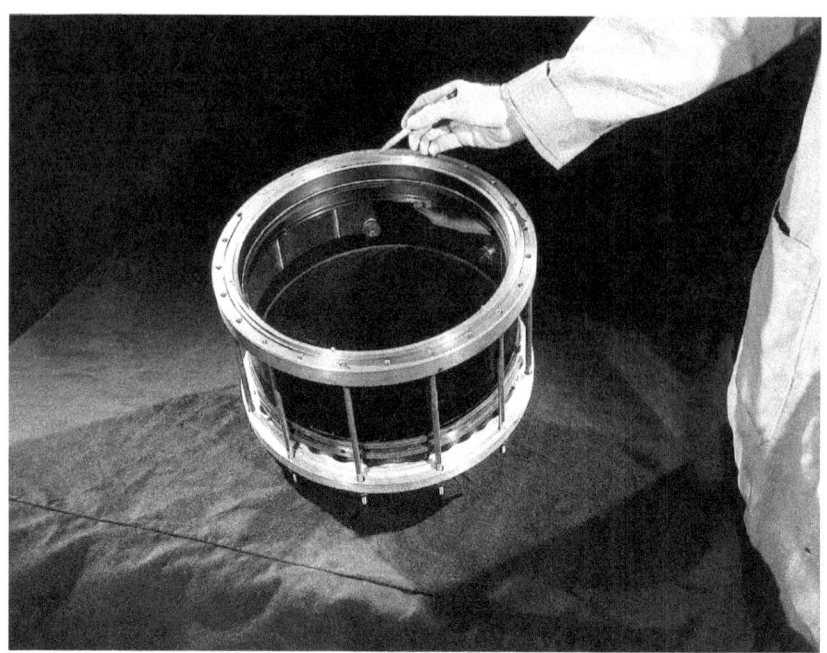

The Wilson Cloud Chamber at AEC's Brookhaven National Laboratory is used for cosmic ray studies. Image courtesy of Wikipedia Commons.

Trails from alpha particles (short & thick) and beta particles (long and thin). Image courtesy of Wikpedia Commons.

The cloud chamber replicates the conditions in our upper atmosphere, most of the air it contains is pumped out, it is chilled and a source of alpha and/or beta particles is placed next to it. The energetic particles cause condensation of tiny droplets of super-cooled water vapour which look like the contrails from jet aircraft. Once these droplets are formed clouds rapidly form as more water vapour condenses out which block sunlight and so cool the Earth.

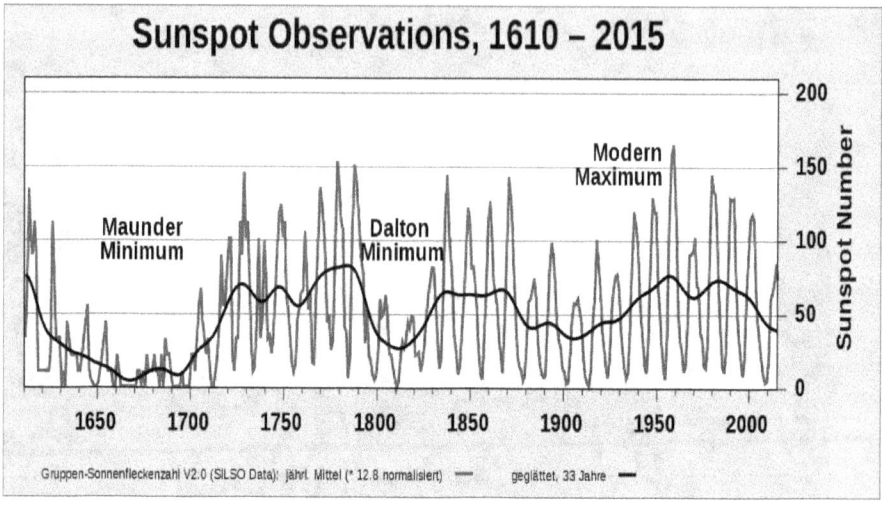

Image courtesy of Wikipedia Commons.

The Maunder, and to a lesser extent the Dalton Minimae, caused the Earth to cool. It was the severe weather of the Dalton Minimum that inspired all the 19th Century's depictions of snowy Christmases. The Maunder Minimum in the second half of the 17th century caused the River Thames to freeze in winter with ice two feet thick and Frost Fairs were held on the river. The graph below shows that the output of the Sun as felt on Earth has gone from a fall of 0.6 Watts per square metre of the Earth's surface at the Maunder Minimum, to an increase of 0.25 WM2 in the late 18th century to fall once again in the early 19th century with the Dalton Minimum. The trend in the 20th century was a rise of 0.6Wm2 which has persisted and levelled off in the 21st century.

Solar irradiance is the **power** per unit area received by Earth from the Sun in the form of **electromagnetic radiation** in the **wavelength** range of the measuring instrument.

The graph below shows the actual heat energy emitted from the Sun to the Earth. The emergence from the depths of the Maunder Minimum and the complete Dalton Minimum can clearly be seen, as can the steady rise from the beginning of the last century.

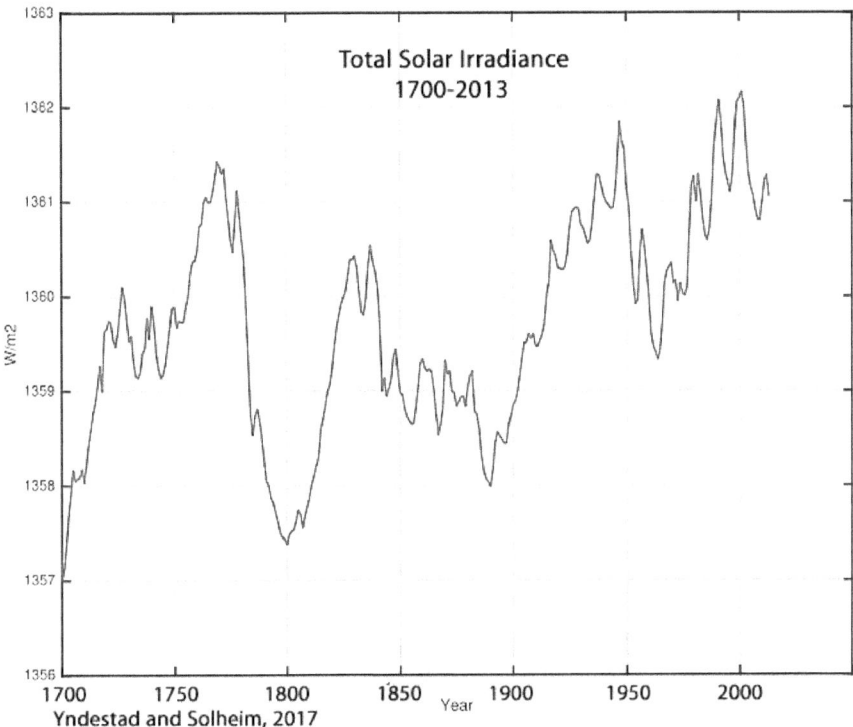

Fig. 3. TSI-HS total solar irradiance from 1700 to 2013 A.D. (Scafetta and Willson, 2014).

Image courtesy of notrickszone.com

This warming phase can be seen to have slowed down from about 1970, the so-called climate 'experts' during the 1970's were telling us about global cooling and how the Earth was going to enter a new ice age. This 'Chicken Licken' mentality is nothing new it has been a 'con'sistent feature of modern 'science'.

Variations in the Earth's climate have been successfully explained and predicted by Milutin Milankovic by the precession of the Earth's Axis (the phenomena where the speed of rotation of the Earth is like a child's spinning top, the 'poles' slowly rotate together other factors into these Milankovitch Cycles. He formulated this theory in the 1920's unsurprisingly carbon dioxide was not then considered to be a problem!

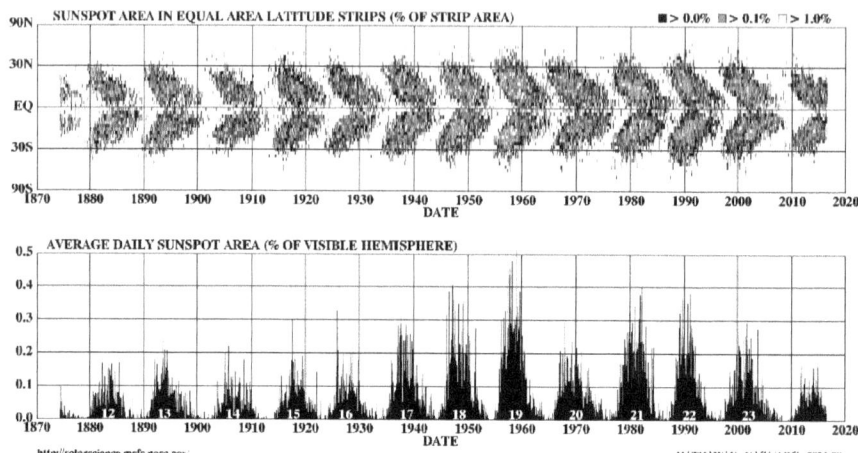

Image courtesy of Wikipedia Commons.

The Sun with a negligible number of sunspots photographed in the violet part of the spectrum courtesy of Wikipedia Commons and Radoslaw Ziomber.

It shows two sunspots only, which has been consistent with Solar Cycle 24 with 268 spotless days at its beginning in 2008. As Solar Cycle 25 begins in 2019 with 132 spotless days up to the end of August, it looks like this trend will continue.

The Evidence.

Much of the evidence comes from weather stations that measure temperature, air pressure, wind speed and direction, humidity and precipitation. Except for those instruments requiring direct exposure to the elements (wind speed and direction, rain gauge), the instruments are sheltered in a vented box, called a **Stevenson Screen**, to keep direct sunlight off the thermometer and wind off the hygrometer (instrument for measuring humidity).

Weather station in a Stevenson Screen courtesy of Wikipedia Commons and Bidgee.

The Stevenson Screen is painted white to reflect heat that would otherwise give false readings. However if the pain peels or gets covered in dust, it will give inaccurately high temperature readings. I could be cynical and say that negating this damage will not be carried out; but I won't!

There is a weather station situated at Heathrow Airport where it is subject to the urban island heat effect of the tarmac runways plus the hot exhausts of full powered jet engines on take-off and reverse thrust assistance when landing. If this goes against common sense, which of course it does, the pinnacle of lack of common sense was reached when a Newcastle City Council worker made Newcastle upon Tyne briefly the most polluted city in the UK. I say briefly, because when the instrument measuring air pollution was moved out of the Eldon Square underground bus concourse, which has diesel fuelled buses entering and leaving every few seconds, it became one of the cleanest cities in the UK.

The most accurate temperature measuring devices are in satellites in Earth orbit that measure atmospheric temperatures by using microwave sensors to measure the temperature of atmospheric oxygen molecules. Broadly speaking they show less warming than the ground thermometers which also have 'adjustments' made, contemporary records upwards, historical ones downwards. These are the easiest of records to falsify since adjustments of less than a two thousandth of one degree per day will equate to a warming of approximately one tenth of a degree per annum. Since their basic premise for carbon dioxide induced global warming is no longer valid, these adjustments, temperatures and the alarmists can be ignored.

Liberalism becomes Populism.

There has been a huge worldwide political change over the last few years. The Union of Soviet Socialist Republics is no more and as I have described earlier, the number of overtly communist countries in the world has gone from 27 outside of the USSR plus 15 within it, to just five. Russia under Putin is once again acting belligerently and President Xi Jinping has returned communist China into the repressive state that it was under Mao who in the country's 'Great Leap Forward' cost 45 million lives over four years, the greatest mass murder in the world's history. The prime and very sad fact is that the Left can only gain and then stay in power by violence and coercion because, as I cannot stress highly enough their ethos is an anathema to basic human instincts of family first and self-preservation second. In the rest of the world Donald Trump, Brexit and further disquiet about the EU from Italy and Greece demonstrate that many people think that big, Left-leaning and liberal government is not the way forward. This process will accelerate when people realise they have been collectively conned by their governments into paying higher taxes on cars, fuel and other higher energy costs based on the untruth that mankind is changing Earth's climate.

Chapter 13: Fake News, Fake Science a Short (Not-faked) History, the Future of Science.

Fake news is not a new phenomenon. In 1693 a printer named William Anderton was tried in the high court and found guilty of High Treason by trying to incite a rebellion against King William III by composing, printing and publishing 'Malicious, Scandalous and Traitorous Libels' about him in an attempt to get him replaced by the Jacobite, King James He was found guilty and executed. William jointly ruled with his wife Queen Mary II until she died of smallpox the following year.

On 21/08/1835 the New York Sun published a series of articles claimin that the British astronomer, Sir John Herschel had discovered life on the moon with using a new high powered telescope. They gave descriptions of bizarre creatures with equally bizarre habits. Sir John was out of the country in a remote place carrying out research and so was unaware of these articles.

A drawing published in the New York Sun in 1835 of the life-forms 'seen' on the Moon.

The New York Sun carried out this hoax to boost circulation which it did, very well in fact. A month later they very quietly announced that it was a hoax.

Nowadays of course we have social media and fake news can spread globally in literally seconds and can worryingly be believed by millions of people. Gullibility concerning news is bad enough, but with science and especially lifestyle it can be catastrophic as I have described previously.

Fake Science.

Correct scientific procedure is based on formulating a hypothesis (the scientist guesses how some physical process works), experiments are then carried out to see if the hypothesis is plausible. If the experiments show it is, it can then become a theory, if the theory is proved to be correct it becomes a Law. The much under-rated physicist Richard Feynman coined this definition, he went on to say it does not matter how smart the scientist is, how clever the hypothesis is, if only one experiment disproves it, then the hypothesis is wrong.

For most people fake science is extremely difficult to identify, because not only is there fake science, there is pseudo-science (pseudo – false). There is a very subtle difference between these terms. Pseudoscience is statements, beliefs, or practices that are claimed to be both scientific and factual, but does not use scientific methodology. It's usually health and/or welfare based. Astrology, Homeopathy, Phrenology and Pyramidology are all examples of pseudoscience.

Fake Science is about advancing a hidden agenda to make societal changes or to make someone or some organisation money and/or credit and fame that they are not entitled to. Climate change hysteria does all of those things and yes, it keeps on giving, especially when it is regarded as being responsible for all the world's ills and those who govern us don't seem to have the first clue about basic science and logic.

Apart from AGW, the other relatively recent fake science theory was that the combined vaccine for measles, mumps and rubella (MMR) given in childhood can cause autism (I have mentioned this earlier with regard to my son). The study was carried out by Dr Andrew Wakefield and others and gained credence by its publication in the Lancet in 1998. Wakefield was subsequently erased from the UK Medical Register. Wakefield has made a great deal of money from this and even worse caused untold harm to many children. He lectures and attracts large audiences in the USA because he and others still believe it to be true.

The classical fake science technique to make societal changes is to use meta-analysis. This is
the process or technique of synthesising research results by using a statistical method to retrieve, select, and
combine results from previous separate but related studies Its most famed usage was to 'prove' that second-hand tobacco smoke was responsible for causing lung cancer in non-smokers.

There are three types of lung cancer: Squamous Cell Carcinoma, Oat Cell Carcinoma and Small Cell Carcinoma. The latter is found in smokers and non-smokers in equal numbers in the periphery of the lungs, the first two are only found in smokers and are found in the bronchus, (the tissue at the centre of the lungs, where the carcinogenic tar from cigarettes is at its most concentrated). When the initial studies were carried out so called 'passive smoking' was found not to be a risk factor for non-smokers. But by blurring the distinction between these different cancers using meta-analysis, the falsehood that non-smokers were at risk of a fatal disease from smokers meant that an indoor smoking ban in public buildings was put on the UK statute books in 2007. Some may think this was justifiable. I do not because it sets a precedent, the thin edge of a very long wedge, allowing politicising of science.

Another example of fake science and social manipulation was the introduction of governmental advice concerning the consumption of alcohol. (Conjoining; 'govern, mental and advice' produces a good description as to the accuracy of this particular guidance and probably many more health advice blunders).

In 1979 the then Minister of Health in the UK government asked the senior medical officer to provide information on safe levels of alcohol consumption. The medical officer said that a study needed to be initiated to correlate peoples drinking habits over 25 years with disease experience. The Minister said that he needed to know now and waiting 25 years was not an option. The result was a guess and to cloud the issue even further 'Units' of alcohol were the measure that people had to use to assess their consumption. 56 units per week was the guess that was made, that was reduced to 36, then to 28, later on to 21 and then to 14, all without a single study being carried out. (GPbG; (Governmental Policy by Guesswork).

Of course adjusting temperatures, describing weather events as climate change is easy, unless their definitions are strictly adhered to, which in the new Faith, they most certainly aren't.

The definition of 'weather': *The state of the atmosphere at a particular place and time as regards heat, cloudiness, dryness, sunshine, wind, rain, etc.*
'If the weather's good, we can go out for a walk?'

The definition of 'climate': *The weather conditions prevalent in an area and in general over a lengthy period.*
'Our cold, wet climate'

Note: Weather and climate are two totally separate concepts, the former is temporary, the latter is semi-permanent and neither is in perpetuity!

"If you can't blind them with science, baffle them with bullsh*t" – WC Fields.

In my view this whole debacle of using science as a social engineering tool has destroyed credibility of all scientists, sadly even the honest ones. Science is about logical investigation, not to bring about worldwide political change by obfuscation, lies and deceit, that is the remit of politicians!

Politics and science are mutually incompatible and should always remain so, sadly but predictably and financially this is no longer the case. The problem is that once one branch of science is abused, the temptation exists for other scientists to do the same. The 'research' and the 'solutions' have cost the taxpayers worldwide billions. The documentary film: *'An Inconvenient Truth'* (whose name I have shamelessly parodied as the title for this book) was released in 2006 and features ex-US Vice-President, Al Gore presenting his views on why and how mankind is causing the planet to warm uncontrollably. At about the same time the fictional film *'The Day after Tomorrow'* was also released which did the same even more dramatically. In 2007 Gore was awarded the Nobel Peace Prize for 'An Inconvenient Truth.' His current financial worth is $300 million, like all socialists it is other peoples' money that should be spent solving the planets' imaginary woes, not his. Likewise he and his cronies can take subsidised flights all over the world telling lies at pointless conferences, but Joe Public must pay a greenhouse gas tax on his annual return flight for his well deserved holiday.

The Climate Policy Initiative webpage in 2013 bemoans the following: 'COPENHAGEN-Global investment in climate change plateaued at USD $359 billion in 2012, roughly the same as the previous year, according to a new Climate Policy Initiative (CPI) study. Once again the figure falls short of what's needed. The International Energy Agency projects that an additional 'investment' of $5 trillion is required by 2020 for clean energy alone to limit warming to two degrees Celsius'. How much more do they want? This money could be used to educate those in countries where abject poverty exists, to provide clean water, eradicate preventable disease and feed the malnourished and starving. Instead it is being used to prop up vanity projects in the West that will impoverish us too. No-one wins, especially not the 'usual suspects' ie the poor souls who live on the breadline in Third World countries.

This is why I loathe the Left with a passion the Left are only interested in penalising the wealthy, not helping the poor. They cannot comprehend that applying disincentives to wealth creation affect everyone, rich or poor. The Left believe that they occupy the moral high ground making questioning of their motives and actions a perceived heresy. They also fail to learn from history, as was shown when Gordon Brown was Prime Minister and he brought in new higher rates of income tax just over 30 years after his predecessor tried exactly the same thing. Predictably income tax revenue fell as they did under a previous Labour government.

Albert Einstein famously said: 'The definition of insanity is repeating the same series of actions and being surprised that there was no different outcome.'

Winston Churchill: 'Capitalism is the unequal sharing of rewards, Socialism is the equal sharing of miseries'.

Grants and Greed the Need for Change.

As I have already mentioned, the global warming gravy train has made some people undeservedly very rich, and even more undeservedly respected. Fraud is the word that springs to my mind! We have all been paying over the odds for fuel for cars, energy for homes and anything else that produces carbon dioxide . Even the word 'emissions' is emotive due to its negative connotations this is despite the fact that all life on Earth (including plants at night), 'emit' carbon dioxide. Animals of course don't emit it, we and they, exhale it. Green leaved plants produce oxygen during the day, they don't emit it. Science is supposed to be logical and factual, not political; apparently climate 'science' is different, but why? Probably, because the outcome of any climate research is pre-determined. Scientific research is peer-reviewed, in other words a scientist or group of scientists carry out research their research is then reviewed by a separate group of scientists and if they find the methodology has been correctly followed and the conclusion is sound, the paper can then be published. The problem is that when a belief is involved it will cloud the judgement of the peers doing the reviewing. In other words any study that shows that the climate is not warming is going to fail the review process by the bias of the reviewers. This is why the concept of AGW is self-perpetuating.

The 'scientists', by making statements like '97% of climate scientist agree that the climate is warming' give themselves away. Science is not and never has been about consensus, scientific procedure is to continue to study and experiment in an attempt to prove or disprove something. The only exception to this process is with a scientific Law where the evidence has been rigorously tried and tested and passed all the tests and trials. Likewise the statement; 'the science is settled' is nonsensical, because there is no law of anthropogenic climate change. The scientific methodology of conjecture, hypothesis, experiment, theory and then law, have served mankind well for many centuries and are still fit for purpose. The fact that a lot of taxpayers' money has been thrown at this 'problem' is at its very root, the whole way science is funded needs to be changed as does the cronyism that currently exists. I do not know enough about it to suggest a change but it clearly isn't working as demonstrated with climate science, and as I mentioned previously, people's health.

What Next?

As I have said previously about RNA and DNA and the possibility of these molecules being unique to Earth, we may very well be the only planet in the universe with sentient and non-sentient life. The Sun will not let the Earth be habitable forever; we need to look after our planet and its inhabitants to ensure the most basic instinct of all sentient life, survival. At some time in the distant future we will have to find other planets to colonise, expensive and intermittent energy will not provide the means to do that that, neither will polluting this world with plastics, toxic and corrosive chemicals. Should mankind colonise other planets in the distant future I would hope lessons would be learned and acted upon with regard to pollution and politics.

Carbon dioxide is not a pollutant, despite the alleged horrors that it supposedly causes, higher concentrations will be beneficial to plants and animals as they were in the past and to humans too in the future. There is no logic to the 'Green' energy policy which is simply a propaganda tool. Seeing wind turbines gently rotating 24 hours a day whether the wind is or isn't blowing sends out the subliminal message of reminding people about global warming and reinforcing the idea that our prosperous societies need to change for the worse.

The ultimate source of energy is hydrogen fusion, the process that makes the stars and our Sun shine. The technology to deliver this energy is some decades away therefore in the short term we need to use fossil fuels with the added advantage that the extra atmospheric carbon dioxide will be beneficial for plant growth and increasing the food supply. In the medium term Thorium Nuclear Reactors are the way forward. One ton of thorium provides the same amount of electrical energy as 3.5 million tons of coal, there is enough thorium in the USA alone to satisfy their energy requirements for a thousand years.

Chapter 14: A Depressing Conclusion.

Most people think that fascist parties of the 1930's were from the political Right, they were not, they were from the political Left, although Mussolini's political party (Partito Nazionale Fascista) were not Marxist, they wanted an economic system in which the employer and the employee were united in associations to collectively represent the nation's economic producers and work together with the state to establish a national economic policy.

The Nazi Party was Left leaning (National Socialist German Worker's Party), which replaced the German Worker's Party its policies were based on Mein Kampf, the book that the Nazi party's policies were based upon. The fascists hated the communists as they regarded them as being too far to the political left they also hated the Jews and thought both these groups wanted to dominate the world. He made bosses and owners of companies sit with their workers at meal times and work for the common good, which was one of his better ideas. Hitler introduced the concept of the Volkswagen (peoples' car) to bring affordable private car ownership to the working people. This process was started in 1933 when Hitler came to power, he appointed Ferdinand Porsche to take charge of this project. By paying 5 Marks a week Germans could save in a government scheme to buy their own car. The 'Beetle' as it was affectionately called was the biggest selling car ever. In 1972 having been in continuous production since 1938, it outsold the previous record holder the Model T Ford. Volkswagen is currently the world's biggest motor manufacturer.

The current obsession with climate change is the same sort of unifying force that led to the rise of the Socialist Left in the last century. The difference is hatred of the Jews then and climate change now, both are wrong, the German Jews were not trying to install a communist German government, which is what Hitler accused them of, this was just an excuse to persecute them firstly and secondly to unite German citizens after the humiliation of losing the First World War. Climate change is nothing to do with mankind it is a completely natural process portrayed as man-made. Manipulation by using a falsehood for the pursuit of social engineering is something the Left are adept at, because as I have already said, and I make no apologies for repeating it. Human and other animal's brains are hard-wired by millions of years of evolution, the immediate family come first and when the efforts by the breadwinners to provide for them are thwarted, they rebel. These instincts cannot easily be over-ridden.

'Life is not fair' is a truism. No, life isn't fair, it never will be; deal with it! Some people have disabilities, physical, mental or both with a varying range of severity. Others contract disease, where their life-expectancy is greatly reduced some people are intelligent others aren't. The list is endless. Those who are good at creating wealth (at their own risk) should be encouraged to do so, not aggressively penalised and treated as social pariahs. Sadly the 'life isn't fair mentality' is taking over. The millennial generation don't want to hear anything that may make them feel uncomfortable and need their 'safe spaces' so debate is stifled and as a consequence so is exchange of opinion. We have entered a societal phase of mediocrity people cannot think for themselves any more, individuality is frowned upon, needless regulations have replaced common sense and innovation has been replaced by conformity. These attributes occur in phases and probably the last time society was this dour was during Oliver Cromwell's Protectorate after the execution of Charles I in 1645. His Protectorate lasted for five years until his death.

Appendices: It was my intention to include UN Agenda 2030 as an appendix but the publisher' plagiarism checker ruled that out. My advice is to read it and form your own judgement. The hyperlink I have copied again at the end of Appendix 1, together with the motherboard link.

Appendix 1: A Summary.

1) For this first appendix I want to bring together all the evidence demonstrating that increased atmospheric carbon dioxide concentrations only have a minimal effect on Earth's climate and why they can be ignored. The concentration of atmospheric carbon dioxide in the early days of Earth was 20% compared with 0.04% now. The Sun was 30% cooler than it is now but it defies all logic that increased carbon dioxide atmospheric concentrations are going to cause catastrophic warming in the near future when they are 1/500 of what they were in the past. It would have been impossible for life to begin and then evolve if atmospheric carbon dioxide can cause significant, let alone catastrophic climate change.

2) Oxygen only exists in large quantities in or atmosphere because of photosynthesis, there is no other natural method of its production, photosynthesis can only occur between 0 Celsius and 40 Celsius; biochemistry will not allow it to occur outside these temperature ranges. Photosynthesis is at its most efficient between 10 Celsius and 20 Celsius and needs light, so it cannot take place efficiently at the North and South Poles for the majority of the year. It is at its most efficient between the Tropics of Cancer in the Northern Hemisphere and Capricorn in the Southern Hemisphere where it will occur every day of the year even with moderate cloud cover. The implication for this must be that between these latitudes the primordial temperature must have been between 10 and 20 Celsius. At the equator the average temperature is currently 31 Celsius, so with 30% less solar output all those millennia ago, that would put the equatorial, primordial temperatures at about the optimal level for photosynthesis to occur.

3) Not all wavelengths of infra-red radiation (heat) are absorbed by carbon dioxide molecules, the vast majority allow the

infra-red photons to leave the Earth totally unimpeded. Those photons that are absorbed by a carbon dioxide molecule are absorbed once only and re-emitted in a different wavelength. Any carbon dioxide molecule higher up in the atmosphere is less likely to interact with an infra-red photon than one nearer the ground. That is why a higher concentration of more than 20 parts per million has very little effect on atmospheric temperatures, because this effect is logarithmic, not linear.

4) Photons hitting carbon dioxide molecules and as a result producing heat can only occur close to the Earth's surface. I have tried to find studies that demonstrate global warming at increased altitudes. The only reference I could find for over 30,000 feet was that climate change will make flights more unpleasant due to an increase in high altitude turbulence. Is it not strange how every aspect of warming has something negative associated with it? Various 'studies' have been carried out that claim that higher altitudes will suffer more from climate change than lower ones. They will probably have the same 'corrections' made to the historical data (downwards) as Stevenson screened weather stations have, placed near to roads that are usually warmer than grassland or woodland which are diligently recorded as accurate.

5) The Sun is the main driver of our climate not carbon dioxide, the link between lack of sunspots and global cooling is clear, observations of the Sun have been carried out since the invention of the telescope over 400 years ago. We could be entering a phase of global cooling with increased cloud cover, so solar cells will generate less electricity and since there would be less energy available, winds will not be as strong and the oceans will cool, so global rainfall will be diminished, making all the renewable energy sources wholly inadequate.

6) The solutions to the alleged problem are illogical for two reasons. Wind, solar, tidal and hydro cannot provide enough energy reliably and economically for a 21st century computer

controlled world. The amount of carbon dioxide produced in their manufacture, implementation and maintenance is probably the same as burning fossil fuels to achieve the same result. Fusion, when the technology is developed can, Thorium technology can do so now, with no risk of reactor meltdowns or radiation leaks.

7) Every single prediction about catastrophic climate change without exception has been wrong, Arctic sea ice extends more or less to where it normally extends to and is not ice free as it was supposed to be a decade ago, southern Spain is not a desert and hurricanes and typhoons are less common as opposed to being an ever increasing threat. Crop failures and riots that were predicted to have happened 20 years ago are still not occurring and in my view, never will. The only climate refugees are those moving closer to the equator, not as we have 'reliably' informed, away from it.

https://motherboard.vice.com/en_us/article/43pek3/scientists-warn-the-un-of-capitalisms-imminent-demise

https://sustainabledevelopment.un.org/post2015/transformingourworld

Appendix 2: The Main Driver of Our Climate.

The main driver of the Earth's climate is of course the Sun it is most certainly not carbon dioxide as some people expect us to believe.

The Sun is about half way through the most stable part of its life. Over the course of the past four billion years, during which time planet Earth and the entire Solar System were born, it has remained relatively unchanged. This will remain the case for another four billion years, at which point, it will have exhausted its supply of hydrogen fuel. So far, the Sun has converted hydrogen of an estimated 100 times the mass of the Earth into helium and solar energy. As more hydrogen is converted into helium, the core continues to shrink, allowing the outer layers of the Sun to move closer to the centre and experience a stronger gravitational force. This means there is more pressure on the core, which is resisted by a resulting increase in the rate at which fusion occurs. Basically, this means that as the Sun continues to expend hydrogen in its core, the fusion process speeds up and the output of the Sun increases. At present, this is leading to a 0.01% increase in luminosity every million years, resulting in a 30% increase over the course of the last 4.5 billion years. I concede that this is a source of global warming, but a source we have absolutely no control over whatsoever even with our current and most probably our future technology.

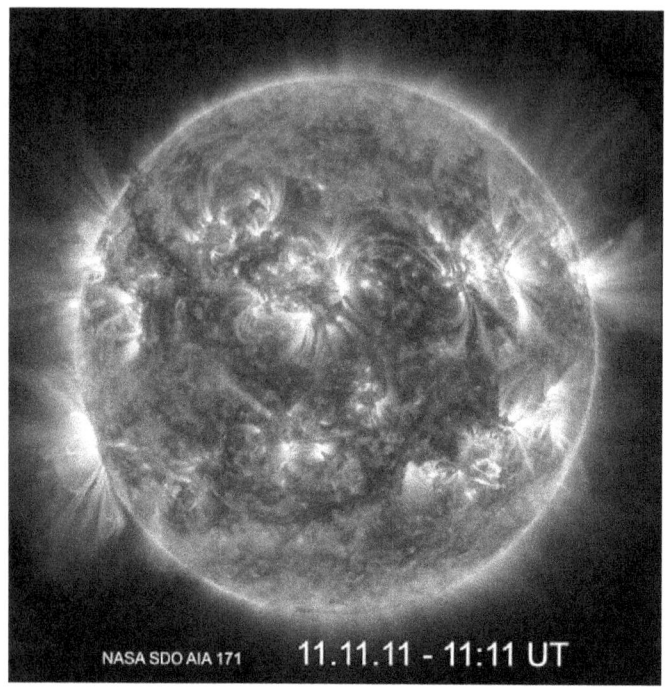

The Sun photographed by NASA's Solar Dynamics Observatory Spacecraft.

In 1.1 billion years from now, the Sun will be 10% brighter than it is today in 3.5 billion years from now, the Sun will be 40% brighter than it is now. This increase will cause the oceans to boil, the ice caps to permanently melt, and all water vapour in the atmosphere to be lost to space. Under these conditions, life as we know it will be unable to survive anywhere on Earth.

In the distant future the Sun will run out of hydrogen fuel, this will begin in approximately 5.4 billion years. With its hydrogen exhausted in the core, the inert helium ash that has built up there will become unstable and collapse under its own weight. This will cause the core to heat up and become much denser causing the Sun to grow in size and enter the Red Giant phase of its evolution. It is calculated that the expanding

Sun will grow large enough to envelop the orbits of Mercury, Venus, and probably Earth. Even if the Earth survives, the intense heat from the red sun will scorch what is left of our planet.

Once it reaches the Red-Giant phase of its life, the Sun will have approximately 120 million years of active life left. But much will happen during this amount of time. First, the core (full of helium), will ignite violently (called a helium flash); where approximately 6% of the core and 40% of the Sun's mass will be converted into carbon within a matter of minutes. The Sun will then shrink to around 10 times its current size and 50 times its current luminosity, with a temperature a little lower than it is today. For the next 100 million years, it will continue to fuse helium in its core into heavier elements until the helium is exhausted. Then it will expand again (much faster this time) and become more luminous.

Over the course of the next 20 million years, the Sun will then become unstable and begin losing mass through a series of thermal pulses. These pulses will occur every 100,000 years +/-, becoming larger each time and increasing the Sun's luminosity to about 5000 times its current brightness and its radius to over 1AU (an Astronomical Unit is 93 million miles, the distance from the Earth to the Sun). At this point, the Sun's expansion will envelop the Earth, leaving it a dry, barren, airless rock. Planets in the Outer Solar System will change dramatically, as more energy is absorbed from the Sun, causing their water ices to sublimate, perhaps forming dense atmospheres and surface oceans. After 500,000 years or so, only half of the Sun's current mass will remain and its outer layer will begin to form a planetary nebula (a ball of gas that looks like a planet when viewed through a telescope).

This period of its evolution will be even faster, as the ejected mass becomes ionised to form a planetary nebula and the exposed core reaches 30,000 K. The final, naked core temperature will be over 100,000 K, after which the remnant will cool towards a white dwarf where no nuclear reactions take place. The planetary nebula will disperse in about 10,000 years, but the white dwarf will survive for trillions of years before fading to become a cold black dwarf.

Appendix 3: The Evolution of Man.

Life is classified by the following twelve levels: Domains, Kingdoms, Phyla, Class, Order, Family, Sub-Family, Tribe, Sub-Tribe, Genus, Species and Sub-Species. Most living things are identified by their genus, species and sub-species so a human is Homo sapiens sapiens. The diversity of life on Earth is such that in May 2016, scientists reported that 1 trillion species are estimated to be on Earth currently with only one-thousandth of one percent catalogued.

The Family; Hominidae contains all apes chimpanzees monkeys and man it excludes extinct species. The Sub-Family; Hominine includes gorillas, chimpanzees and man. The Tribe has just chimpanzees and man as members. The Sub-Tribe; Homininan has modern humans and extinct close relatives such as the australopithecines, but excludes chimpanzees The genus; Australopithecus were the ancestors who evolved into the Homo genus, which is the direct line of evolution to modern man. The australopithecines were intermediate between apes and humans, australopithecines and humans are biologically very similar we hominans share the fact that we evolved from the same ape ancestors in Africa, that both genera are habitually bipeds, or two-footed, upright walkers. By comparison, chimpanzees and gorillas are primarily quadrupeds, or four-footed.

There have been several important fossil discoveries in the African continent of what may be very early ape/hominins, also known as proto-hominins. These creatures lived just after they had diverged from our common hominid ancestor, thought to be chimpanzees. It is thought the earliest australopithecines most likely did not evolve until around 5 million years ago or shortly afterwards, during the beginning of the Pliocene Epoch in Eastern Africa. The primate fossil record for this crucial transitional period leading to the emergence of the australopithecines is still scanty and somewhat confusing. However, by about 4.2 million years ago, australopithecines were definitely present. The point at which the genus Homo evolved is Homo habilis which

evolved around 2.8 million years ago. Homo habilis is the first species for which we have definitive evidence of their use of stone tools. About 1.9 million years ago Homo erectus first appeared in the fossil record, with a cranial capacity that had doubled. Homo erectus was the first of the hominans to emigrate from Africa, and,
from 1.8 to 1.3 million years ago, this species spread through Africa, Asia, and Europe. One population of Homo erectus is also sometimes classified as a separate species (Homo ergaster) which remained in Africa and eventually evolved into Homo sapiens (modern man). It is believed that both these species (Homo erectus and Homo ergaster), were the first to use fire and relatively complex tools.

The earliest transitional fossils between Homo ergaster/erectus and archaic Homo sapiens sapiens, are from Africa, such as Homo rhodesiensis. Seemingly transitional forms were also found in what later became the country of Georgia. These descendants of African Homo erectus spread through Eurasia from about 500,000 years ago, evolving into Homo antecessor, Homo heidelbergensis and Homo neanderthalensis. The earliest fossils found of anatomically modern humans are from the Middle Paleolithic Period, about 200,000 years ago. Later fossils from Israel and Southern Europe were discovered from around 90,000 years ago.

As modern humans spread out from Africa into the rest of the world they encountered other hominans such as Homo neanderthalensis whose evolution had gone down a different path due to their genetic need to live in different conditions. The nature of interaction between early humans and these sibling species has been a long-standing source of controversy, the question being whether humans replaced these earlier species or whether they were in fact genetically similar enough to interbreed and produce viable offspring who could also reproduce. In which case, those earlier populations probably contributed genetic material to the genome present in Homo sapiens sapiens. The problem with inter-species reproduction is that usually the offspring are sterile. The most well known example of this is the mule, where one parent is a horse and the other is a donkey.

This migration out of Africa is estimated to have begun about 70,000 BC and modern humans subsequently spread globally, replacing earlier hominins either through competition or hybridisation. They inhabited Eurasia and Oceania by 40,000 BC, and the Americas by at least 14,500 BC.

Appendix 4: DNA & RNA.

Comparison of DNA and RNA

Comparison	DNA	RNA
Name	Deoxyribo-Nucleic Acid	Ribo-Nucleic Acid
Function	The long-term storage of genetic information and transmission of genetic information to make new cells and new life-forms.	Used to transfer the genetic code from the nucleus to the ribosomes to manufacture proteins. RNA is used to transmit genetic information in some life-forms and may have been the molecule used to store genetic codes in primitive life.
Structural Features	B-form double helix. DNA is a double-stranded molecule consisting of very long chains of nucleotides.	A-form helix. RNA usually is a single-strand molecule consisting of shorter chains of nucleotides.
Composition of Bases and Sugars	Deoxyribose sugar with a phosphate backbone based on adenine, guanine, cytosine, thymine bases	Ribose sugar with a phosphate backbone based on adenine, guanine, cytosine, uracil bases
Propagation	DNA is self-replicating.	RNA is synthesised from DNA on an as-needed basis.
Base Pairing	AT (adenine-thymine) GC (guanine-cytosine)	AU (adenine-uracil) GC (guanine-cytosine)

Reactivity	The C-H bonds in DNA make it relatively stable the body destroys enzymes that would destroy DNA. The small grooves in DNA also serve as protection, providing little space for enzymes to become attached.	The O-H bond in the ribose of RNA make the molecule more reactive, compared with DNA. RNA is unstable under alkaline conditions, the larger grooves in the molecule make it more susceptible to enzyme attachment. RNA is constantly manufactured, used then degraded.
Ultraviolet Damage	DNA is susceptible to ultra-violet light damage.	Compared with DNA, RNA is relatively resistant to ultra-violet light.

Both DNA and RNA are the only molecules that have a different chemical combination of atoms for each individual life-form. There are an infinite number of variations of both DNA and RNA. Every other molecule that we are aware of only has one combination of atoms.

Acknowledgements.

Many thanks to my family and friends who had to take a 'back seat' from my attention as I was constantly thinking about the style, format and content of this, my first and depending on the sales figures, probably my last book. It is the message that is important, not the content nor the writing style nor the grammar.

www.ingramcontent.com/pod-product-compliance
Lightning Source LLC
Chambersburg PA
CBHW051325220526
45468CB00004B/1506